For the Love of

hops

The Practical Guide to Aroma, Bitterness and the Culture of Hops

Stan Hieronymus

brewers publications

A Division of the
Brewers Association
Boulder, Colorado

Brewers Publications
A Division of the Brewers Association
PO Box 1679, Boulder, Colorado 80306-1679
BrewersAssociation.org
BrewersPublications.com

Printed in the United States of America.

10 9 8 7 6 5 4 3 2 1

ISBN: 1-938-469-01-1
ISBN-13: 978-1-938469-01-5

Library of Congress Cataloging-in-Publication Data

Hieronymus, Stan.
   For the love of hops : the practical guide to aroma, bitterness, and the culture of hops / by Stan Hieronymus.
       pages cm
   Summary: "Discusses the science and culture of hops, exploring such topics as history, hop varieties, brewing and dry-hopping techniques, and provides commercial recipes for brewing beer"-- Provided by publisher.
   ISBN-13: 978-1-938469-01-5
   ISBN-10: 1-938469-01-1
   1. Hops. 2. Brewing. 3. Beer. I. Title.

   SB317.H64H54 2012
   633.82--dc23
                        2012033604

Publisher: Kristi Switzer
Production & Design Management: Stephanie Johnson Martin
Cover and Interior Design: Julie White
Cover Illustration: Alicia Buelow
Technical Editor: Dr. Christina Schönberger
Copy Editing & Indexing: Daria Labinsky

To all of those who have been scratched by the hops.

Table of Contents

Acknowledgments

In the final paragraphs of this book, David Grinnell of Boston Beer Company observes, "There are people who have given their life to this plant." *For the Love of Hops* would not have made it to print without their help nor without the generosity of many others.

I have no idea why Evan Rail ventured out from Prague to the Czech Republic countryside on a ridiculously cold February day to collect the recipe for *14° Tmavé Speciální Pivo*, but I am grateful, and you should be, too. I could offer scores of similar stories about other people I must thank but will try to be brief.

If Kristi Switzer, publisher at Brewers Publications, hadn't given me the chance, I never would have started the book. If Daria Labinsky, my wife, didn't encourage the pursuit of what some would call frivolous endeavors and offer the copy editing support every writer needs, I never would have finished it.

If Christina Schönberger, Val Peacock, Tom Shellhammer, Tom Nielsen, Peter Darby, Matt Brynildson, and Larry Sidor didn't know so much about hops, I might have given up before there was a Table of Contents. I owe particular thanks to Christina, for a) emailing me information I wasn't able to find elsewhere, b) keeping me up to date on the latest research, and c) providing her unique technical expertise during the editing process.

Of course, I must thank Ken Grossman for writing the foreword, but perhaps even more for creating an atmosphere at Sierra Nevada Brewing that is conducive to discovery and innovation. I spent no more important research time than that at Sierra Nevada. Each book has a moment or two when things come into focus. For this one, the first was when I sat in on a talk titled "Dissecting Hop Aroma in Beer" by Tom Nielsen from Sierra Nevada at the 2008 Craft Brewers Conference, long before I knew I would be writing this book. The second was during a conversation with Anton Lutz and Elisbeth Seigner at Hüll Hop Research Center, when I realized how far into the twenty-first century Lutz was already thinking.

I also appreciate John Harris' summarizing hop selection much better than I could have, and Peter Darby, Martyn Cornell, and Chris Swersey for providing an extra set of eyes and returning an unreasonable number of emails. Gayle Goschie, Jason Perrault, Florian Seitz, and Leslie Roy, in particular, took time to explain the curiosities of hop growing to a Midwesterner who was reared in the middle of corn and soybean fields. Then Ralph Olson, Jim Solberg, Paul Corbett, and almost everybody at Brewers Supply Group did the same with the hop trade. Also, I should thank a thousand brewers who have shown me their hops over the years, but will keep it to seven: Ron Barchet, Yvan de Baets, Vinnie Cilurzo, Dan Carey, Richard Norgrove, Ted Rice, and Eric Toft.

Just as important were people who helped me find information I wouldn't have found otherwise and got me to where I needed to be. Thanks first to Otmar Weingarten, managing director of the Hallertau and German Hop Growers Association, for his hospitality. Also to Ute Lachermeier, Jürgen Weishaupt, Michal Kovarik, Troy Rysewyk, Glenn Payne, John Humphreys, and Rebecca Jennings.

Finally, there was the matter of turning the manuscript, and all its parts, into a book. Thanks, once again, to Stephanie Johnson Martin and Julie White for making that happen.

A few days before the 2011 harvest began I sat on the Tettnang town square with Weishaupt, managing director of *Tettnanger Hopfen,* and listened to a community band play. He later introduced me to one of the clarinet players, a former Tettnang hop queen. Earlier that day he had said, "We like to say we live with the hops."

Now I can say the same. I have to thank my family—Daria, Ryan, and Sierra—for letting me do that.

Foreword

Although I don't have an exact recollection, I have to believe that my first encounter with the aroma of hops probably occurred sometime in my fifth or sixth year of life. While growing up, I spent a lot of time at my neighbor's/good friend's house. His father, Cal, besides being a metallurgical rocket scientist at Rockwell, was an accomplished homebrewer. Throughout my childhood I watched him spend many weekends in the kitchen with large pots on the stove, boiling together what appeared to be mysterious potions from an assortment of exotic ingredients. I was intrigued with his makeshift brewing equipment, which was composed of colanders, hoses, and buckets that he had co-opted and converted from his wife's kitchen and the hardware store, long before such things had been developed for hobbyists. I still vividly recall the occasional and exciting boilover and the unbridled reaction of his generally tolerant wife to the mess on her stove. The unusual aroma of boiling wort captivated me early on. Although some people find it disagreeable, I was mesmerized by the pungent, sweet malt and floral hop aroma. It wasn't like anything I had ever smelled before.

For years growing up, I was regularly around the entire brewing process, from simmering kettle, covered crocks, and plastic buckets with foamy heads, to a row of glass carboys with their airlocks bubbling away on the service porch. When we were older, we would occasionally get tasked to help wash bottles or put caps on during bottling day. Brewing was Cal's hobby, and it soon became my passion. I now understand that this early exposure to hops and brewing had a major influence on my career path, and, later on, the style of beers I chose to brew and enjoy. I have often asked myself the question, "Would I be a brewer today if I hadn't been seduced at an early age by the fragrant hop?" I have been what is now called a hop head for most of my life. Although I enjoy and appreciate a range of beer styles, I always gravitate to the hoppy side and never tire of the complex and aromatic qualities that the wonderful hop has to share.

My neighbor started brewing before quality ingredients, particularly hops, were available for hobbyists. Luckily, he had befriended someone who worked in a large, international brewery and occasionally received "samples" of some of the finest hops from around the globe. They were leaps and bounds better than what was generally available in homebrew supply shops, and had origins in faraway places like England, Yugoslavia, Poland, Germany, and Czechoslovakia. Later, as I became an aspiring homebrewer, I remember how fortunate I was to have access to these generally unknown and unavailable hops and how much they improved the quality of my beer.

If you were a homebrewer in the '60s or '70s, what was being sold for "hops" were generally small, compressed bricks wrapped in pink paper that were simply labeled "hops." At best they contained two- or three-year-old Cluster hops that almost certainly had been in common storage, meaning unrefrigerated, but they may have been any variety that had been rejected by brewers or were in surplus at the time. The limited availability of quality hops was not necessarily related to a conspiracy against homebrewers. Because it was still illegal to brew beer at home, the homebrewing movement was still in its infancy, and few had realized the market potential of supplying hops to hobbyist brewers. Interestingly enough, these pink, 4-ounce bricks of hops were originally destined for Africa and South America to be used in bread baking and had found their way into the homebrew trade in the United States as a sideline. The origin of the use of hops in bread is a bit of a mystery; I have found references to it in old recipes as an ingredient in the yeast starter. Although hops may have been used for added flavor, it's more likely that someone stumbled upon the fact that the antiseptic properties of hops helped keep their unrefrigerated yeast culture from souring. Some years ago, in talking with the supplier of these pink bricks, he told me the story of the year he ran out of his typical aged hops. He was forced to supply fresh hops to his customers, which weren't well received, and they rejected them because of their pungent smell. His solution was to break up the bales and leave them out in the sun for several weeks to age them quickly and drive off any semblance of hoppy aroma.

Largely driven by my inability to find quality hops and other home-brewing supplies in my area for my increasingly serious hobby, I made the decision to open up my own store, Home Brew Shop, in Chico, California, in 1976. That year my fascination with the role that hops played in beer took me on a road trip to Yakima, Washington. I connected with growers at the farmer-owned hop co-op, as well as an energetic young Ralph Olson, just new to the hop trade. Although the Home Brew Shop was too small to buy a typical 200-pound bale of any one variety, I talked them into selling me more than 100 individually wrapped, one-pound "brewers' cuts" of every variety they grew. These are the samples that are removed from every 25 or 50 bales in a lot, and they are normally provided to breweries for quality evaluation. The U.S. hop industry at the time was still primarily focused on just a few varieties, primarily the hearty, reliable, and easy-to-store Cluster. I recall procuring Bullion, Northern Brewer, and Brewer's Gold on my first trip. I later made arrangements to purchase some Cascades and Willamettes from Oregon, since they were not yet established in Washington fields. Cascade, the hop that would become Sierra Nevada's hallmark, had just recently been released. I loaded up my Toyota station wagon and drove home to Chico, ecstatic with my new direct connection. Of course, my excitement was probably heightened by a hop oil-induced buzz from all the freshly harvested cones. Later that year I also established direct connections with European merchants, opening up a whole new world of varieties for homebrewing experimentation.

Although my love affair with the hop makes up just a small piece of the puzzle in the evolution, cultivation, and use of hops, it occurred at a pivotal time in history for this often marginalized but essential brewing ingredient. When I started my commercial brewing career in 1980, hop-forward commercial beers in this country had all but disappeared. There were a handful of hoppy imports and a few remaining relics from an earlier era, such as *Rainier Ale* and the hop oil-infused Ballantine. The sale of these brands was in steep decline, and I witnessed them being reformulated and losing some of their hop edge to make them more appealing to a younger or less discerning audience. In general, beer was heading down a path of homogeneity and blandness, with a general reduction of both malt usage and hop character, with the shift toward lighter and less distinctive lager styles. Hops were relegated to a minor role in most beers, as breweries strove to produce beers that offended no one's taste sensibilities.

Fritz Maytag was probably the first American craft brewer to refocus on hops, followed soon after by the initial wave of craft brewers like me, whose homebrewing roots gave us an appreciation for the increasingly scarce hoppy beers. In recent years craft brewers have taken their love of hops further and turned conventional use and the generally accepted target for "hop aroma" on their heads. Only several years ago, established brewers in both the United States and around the globe eschewed the pungent, atypical aroma that craft brewers now wholeheartedly embrace.

This book is an amazing compendium on the hop, written at a level of detail that will captivate historians, chemists, and brewers alike. Stan Hieronymus' exhaustive research traces the history and evolution of many traditional and recently developed cultivars that have been embraced and supported by craft brewers and reveals a great deal about the dynamics driving the industry. Stan offers personal insight into the multigenerational families that continue to struggle to meet the ever-changing needs of the brewer. This book is technically sound, very well researched and footnoted, and digs into the use and history of hops in a deep and relevant way, for those in the brewing industry and those just curious about this amazing plant.

Ken Grossman, Founder
Sierra Nevada Brewing Co.

Introduction

Hops in the twenty-first century

Every August farmers from across Bavaria's Hallertau hop growing region take an evening off from preparing for the upcoming harvest to join about 3,000 others in a large tent in the center of Wolnzach, where they order plates overflowing with *volkfest* food, listen to a traditional brass band, clink one-liter mugs of beer together, and ultimately vote for the new *Hallertau Hopfenkönigin*. "It is a dream for each hop grower's daughter," said Christina Thalmaier, who won the honor in 2010, 60 years after the first queen was crowned. "And for the father of every daughter it is also a dream."

Hops and beer queens from other parts of Germany are introduced, along with several brewmasters from the region. Representatives of various hop growing organizations dress in their Bavarian best. Each queen candidate must come from a hop farming family and answer a variety of questions, sometimes technical, about hops. Thalmaier said she was still a baby when her mother took her along to work in the family hop gardens. She began trimming and training hops, guiding them clockwise around a string, when she was six years old. "It is work you had to do when you were young. You don't understand why," she said. "When I was young, I hated this work so much."

Veronika Springer, the winner in 2011, helps her family tend to about 40 acres of hops on a farm 15 minutes south of Wolnzach and also works in town at NATECO$_2$, a plant that extracts about 10,000 tons of hops per year, turning pellets into a thick paste packed in tins or drums. It looks not at all like the green bouquet of cones Springer carried with her after she was crowned.

On the evening Springer was chosen queen I circled past NATECO$_2$ during the short walk back to my hotel, Pension Hopfengold. I tried to imagine what the celebration might have been like 40 years earlier, when more than 6,000 farmers in the Hallertau grew hops, as compared to fewer than 1,200 today. The plant, which also handles many other products, was dark, but in a few weeks 85 employees would be at work through three shifts a day, and need more than eight months to process about one-third of the hops grown in the Hallertau.

My day had begun in Žatec in the Czech Republic. Fog covered fields of Saaz hops, and the sun didn't burn through the haze until I was beyond the hop yards and halfway to Pilsen. I took a hop-scenic route, soon driving through hundreds of acres of Hallertau hops and millions of bright green cones on the way to Mainburg, passing more processing plants before turning onto the *Deutsche Hopfenstrasse* leading to Wolnzach.

In a few weeks each acre would yield, on average, eight bales of hops, occupying a space less than 4 by 4 by 8 feet high when stacked side by side. The cones in those bales would later be processed into pellets, then the pellets into extract. The pellets would take up half as much room as the bales, and after extraction an acre of hops might fit in a 45-gallon drum.[1]

These two different forms of hops—one celebrated as a traditional agricultural product that adds to beer a taste unique to the place where it was grown, the other sold as a commodity traded in kilograms of alpha acid—coexist in the twenty-first century. The same is true of hops prized for the aroma, and therefore flavor, they provide beer and hops that supply only bitterness, even at rather low levels. Without such balance the hop business is not sustainable.

In 2011, little more than 20 years after farmers planted them for the first time, high alpha hops accounted for about half the production in the Hallertau. Florian Bogensberger remembers the outcomes his father, Michael, and other Hallertau growers considered before they began planting hops such as Nugget and Target, developed in the United States and England, respectively. The Bogensbergers farm about 350 acres,

160 of them devoted to hops, on property they bought from the Barth family in 1977. "People here were not so sure it was a good idea to have high alpha," Bogensberger said. "If too many ... started producing, it would eliminate others. On the other side, we would not stay relevant in the market." The Hop Research Center at Hüll released Magnum and Taurus, the first German-bred high alpha hops, soon after.

The road between Wolnzach and the farm where Veronika Springer grew up winds through one hop yard after another as it dips into valleys and climbs to the top of hills, occasionally passing through stands of trees that provide yards of natural protection from wind damage. Two days after their neighbor was crowned hop queen, Alexander Feiner and his father, Erwin, made final preparations for harvest. Alexander is a sixth-generation hop farmer who was seven years old when he first drove a tractor during harvest. He was born in 1985, when the Hallertau had more than three times as many hop farmers as today, and is president of the young hop growers association, which has 350 members.

On the road to the Feiner and Springer farms (among others) near Wolnzach.

"It is important for the young growers to get the information, because it is the future," he said. He visited the Yakima Valley in Washington in 2008 and understood immediately that he'd need to return. "I saw these huge farms. You have to know how your biggest competitor thinks about hops," he said.

Perrault Farms, west of Toppenish in the Yakima region, is not huge by local standards but with 750 acres is about six times the size of the Feiners' farm. Jason Perrault, a fourth-generation farmer born in 1971, began winding twine to string hops when he was five years old. He is vice president of sales for the farm and also breeds hops. That gives him a different perspective on the future, one that includes environmentally friendly low-trellis hops, organic hops, and new proprietary varieties that impart flavors and aromas coveted today but considered undesirable when he was born.

"Sustainability has various components," he said. "We're not going to stay in business if every variety we grow is a commodity. Special varieties help sustain the business." About half the hops the Perraults raise are "super alphas." "There's a need for alpha. We have some valuable customers worldwide and we want to serve their needs," Perrault said.

High alpha/bitter hops constitute about 61 percent of hops planted worldwide and produce about 76 percent of alpha acids, which are traded mostly as a commodity. The differences between alpha-rich varieties merit consideration, but ever since hop merchants traveled to the town of Spalt in Franconia more than 500 years ago, buyers have valued hops based on their aroma. *The Barth Report*, which has been published annually with few exceptions since 1877, began differentiating between aroma hops and "bitter value" hops in the 1970s.

Aroma hop acreage worldwide shrank 49 percent between 1991 and 2011. Alpha hop acreage dipped 5 percent, but because farmers grew better-yielding varieties that contained higher percentages of alpha acids, overall alpha production increased 59 percent. Those are not particularly good trends for brewers shopping for aroma varieties, but recent interest in what the 2011 *Barth Report* called "flavor hops" hints at change. These are hops, such as Galaxy from Australia and Simcoe and Citra from the United States, with strong, distinctive qualities considered undesirable only a few decades ago (and, in fact, not universally accepted today).

American craft breweries use these and other hops in considerably larger quantities than other breweries, and the trend has spread to other countries. Although U.S. craft brewers made less than 6 percent of beer sold in 2011, they used about 60 percent of domestically grown aroma hops. "This love craft brewers have for hops refocuses attention on the plant," said Alex Barth, president of hop merchant John I. Haas, which is part of the Barth-Haas Group.

U.S. Craft Brewery Hop Usage Rates (pounds per barrel)				
Brewery production	2007	2008	2009	2010
Less than 2,500 barrels	1.75	1.44	1.19	1.79
2,500-25,000 barrels	1.26	1.42	1.03	1.27
More than 25,000 barrels	0.84	0.85	0.93	0.94
All breweries	0.93	0.92	0.95	0.98

Source: Brewers Association

Barth predicted early in 2012 that before the end of the decade aroma hops might account for half of U.S. acreage, compared to just 30 percent in 2011. "My one stipulation is, it will be more than half if it is to satisfy customer demand," he said. "What I preach to the industry (growers) is you don't want to chase what is hot. You need a balanced portfolio."

Any effort to classify hops based on alpha acid percentages can result in confusion. Brewers understand they must boil hops for an extended period to extract bittering iso-alpha acids, but the essential oils that result in aroma and flavor will mostly be lost in the process. Therefore they add hops late in the process—the "aroma" addition. Somewhere between these bittering and aroma additions they may include one for "flavor." Certain varieties that might be used either at the beginning or the end are sometimes called "dual purpose," although some brewers use so-called aroma hops for bittering or high alpha hops at the finish. Many that fall into Barth's "flavor" category contain more bittering potential than any hop available not much more than 30 years ago.

In the mid-1970s the European Economic Commission considered designating "aroma" and "high alpha" hops. The "aroma" hops would have been defined as low in alpha acids (less than 7 percent) with the composition of their oils "typical" as determined by gas-chromatographic analysis. All remaining varieties would be classified as "high alpha" hops. D. R. J. Laws argued against the proposal in the 1976 annual report from the Department of Hop Research, Wye College. He pointed to British brewers' experience with dry hopping—the addition of a portion of hops after fermentation—and reported they had learned newer high alpha (for the time) varieties often imparted more desirable hop aroma.

He concluded, "All hops which are grown commercially contain both essential oils and alpha acids and are therefore capable of imparting both hop character and bitterness to beer. There is no necessary connection between the amounts of alpha acids in hops and either the total content or composition of the accompanying essential oils. Clearly no useful purpose is served by artificial attempts to classify hops as either 'aroma' or 'copper' hops."[2]

Fuzzy as their definitions may be, the terms are unavoidable. For instance, Tony Redsell, England's best known hop farmer, still refers to Northdown as a "dual purpose" hop. "I'm an aroma hop grower," he said in August 2011, as he was preparing for his sixty-fourth hop harvest.

In 1878 English farmers planted 71,000 acres of hops, mostly in the southeast. Today about 3,500 oast houses—distinctive, often round buildings with a white cone and cowl on top, scattered across the country but most prominent in Kent—recall that time. However, the buildings, once used to dry hops, have been turned into expensive, fashionable homes. Only about 2,500 acres of hops remain, half in the Kent region, where Redsell farms 3,000 acres, 200 of them devoted to hops. "It's hard work, but so what?" he said. "When you've got it in the blood. ... There may be only 50-odd of us left (in England), but we're a hardy bunch."

English hop growers found a different equilibrium than in Germany or the United States. "The alpha market is not one to get into," Redsell said, talking about massive oversupply left from the previous three years. Farmers in England still had 17,000 acres under wire in 1970, when Target was the world's premier high alpha hop. Now hops favored for their aroma quality and used mostly by English brewers account for three-quarters of British production.

"There is demand for beer made from locally grown ingredients," said Chris Daws, who grows hops himself, represents 37 growers in a cooperative, and is sales director for Botanix. "Brewers in Sussex want Sussex hops." Daws sees hops from all sides. He farms a modest eight acres, which would be small even in the Spalt region and unheard of in the Hallertau or Yakima. Botanix is a division of Barth-Haas, the world's largest hop supplier, and specializes in extraction and processing, selling products variously described as advanced or downstream.

Words to Remember

It is impossible to write a book about hops without including words that beer drinkers and even brewers rarely use, such as trans-iso-adhumulone. They will be defined as they appear. However, there are a few that come up more often that should be made clear now.

Alpha acids and iso-alpha acids: Hop cones contain alpha acids, which are not bitter. They are transformed (isomerized) into iso-alpha acids, which are bitter, by boiling.

Cones, whole hops: Unprocessed hops (although they are dried and baled, because otherwise they would rot).

Conventional or traditional hopping: The addition of hops in that whole form or in pellets, which contain the same lupulin and green matter as cones.

Craft beer: This book uses the Brewers Association definitions for craft beer and craft brewery. Those can be found here: www.brewersassociation. org/pages/business-tools/craft-brewing-statistics/craft-brewer-defined.

Form: Beyond cones and simple pellets, brewers have many other choices, including hop extracts and a wide range of advanced products. Each of these is a different form.

Odors: Odors are combinations of compounds that become aromas after being processed in our brains.

Saazer-type hops: Saaz, Spalt Spalter, Tettnanger, and Lublin are so genetically similar that it sometimes makes sense to refer to the group as one.

"You can't divorce yourself from market forces. Brewers are savvy. If there is a substitute they will use something cheaper," he said.

His five children take holiday to help during harvest, and he hopes one day one will want to take over the farm. His great-great-grandfather managed 200 acres of hops and was the first in the region to put his hops under wire. In 2009 his Pilgrim hops won overall champion in the English Hop Competition, and in 2010 his Admiral captured first for the best sample of hops bought primarily for their high alpha acid content.

He appreciates that hops themselves are not the end product. "When you can see a brewer using those hops, using hops you grew to make beer the public likes … " he said. "The bloody things get in your system. Your year revolves around it. … You always want a good rain in July, even if it interferes with holiday."

Redsell still spends two or more days a week walking his hop gardens, looking for potential problems. "The hop to me is my life. The thought of getting up in the morning and not having some hops to look at, that would be terrible," he said. "I tend to think more about the hops than beer. For me, beer is a by-product of hops."

A few weeks later and a couple of days before harvest would begin, Alexander Feiner had much left to do, including worrying about the weather. "Too hot for hops," he said, fiddling with a computer system that would monitor the percentage of moisture remaining in hops during the drying process. "The best computer is still your hand." He touched his thumb to his fingers, squeezing an imaginary hop cone—a motion I saw repeated over and over, as if it is a nervous habit hop farmers are born with. Perrault said whenever he walks through a field he finds himself constantly grabbing cones, breaking them open and smelling them. Same thing.

Feiner talked more about technology, about marketing, and about business practices the other young hop growers needed to consider. He has many ideas. They all begin from his relationship with the hop.

He curled the fingers on his left hand as he began to illustrate, reaching over to his right wrist, running his fingernails up to his elbow and quickly back down to the wrist. He did it again, and then he explained.

"They say whoever is scratched by the hops cannot escape them."

Me to Mirror:
So You Want to Write a Book About Hops?

When did I sense I was in trouble? Maybe when we were talking about an effort in the early 1990s to create a Hop Aroma Unit, and Tom Nielsen at Sierra Nevada Brewing said, "It would take somebody doing a Ph.D." to attempt that today. Or perhaps it was six months later during a discussion with Val Peacock of Hop Solutions Inc. about the source of certain compounds, when he said, "That would make a good topic for a doctoral thesis."

In fact, it was probably a moment between, a late spring day in Milwaukee. David Ryder, MillerCoors vice president for brewing and research, sat at the head of a conference table at the brewery's Technical Center. Patrick Ting, Jay Refling, and Pattie Aron, who like Ryder are experts on both theoretical and practical levels, sat across the table. Troy Rysewyk, manager of pilot brewing, was to my left.

This was a place to ask questions. Those around the table do it regularly. "We never accept what we're told. We all ask silly questions," Ryder said. "There aren't any answers to some of them."

They answered questions. They asked new ones. "You go back 20 years, and people accepted the thinking around hops," Ryder said. That explains why information in recently published texts may be out of date or incomplete. "Research was focused on proving what was known."

Belgian researchers offered what amounted to a quick summary of knowledge circa 2010 in the introduction of a paper assessing the impact of advanced hop products on aroma and bitterness. "Nowadays, the chemistry behind hop alpha acids derived from bitterness is well understood, and as a consequence beer bittering is controllable in brewing practice," Filip Van Opstaele and associates wrote. "This definitely holds true when applying advanced bittering preparations … the brewery is able to prepare beers with a reproducible bitterness and an improved bitterness stability on storage."

Hoppy aroma was not, and is not, nearly as well understood. "This gap can be ascribed to the extreme chemical complexity and varietal dependency of hop essential oil itself, and to modifications and losses of hop oil constituents that take place along the brewing process. Consequently, inconsistent hop aromas represent a serious quality problem in view of a reproducible and sufficiently stable beer flavor."[3]

MillerCoors breweries produce many of their beers using advanced hop products. "The way we look at hops is, we have a toolbox. We don't have a view that using other hop products is cheating," Ryder said. Many brewers, including homebrewers, may never make beer with any form of hop extract, but the chemistry involved and the questions left to be answered are much the same.

"There is a lot going on," Aron said. "A lot of people know a lot of things. Not all of them are talking about what they know. People are chasing the pieces. I don't think anybody has put it all together."

She went to work at MillerCoors after completing her Ph.D. at Oregon State University. She's familiar with a beer culture where "balanced" hop aroma and flavor mean something different than they do at the world's largest breweries. "In Japan, Germany, other places, people are excited about hops, about dry hopping. They want to understand American hops," she said.

This book addresses that interest, but it would be short-sighted to stop there. It is intended first for brewers who buy their hops by the pound, brewers who account for a rather small percentage of the world's beer production and an outsized portion of hop consumption. Those in the hop industry appreciate the attention that new, bold varieties cast on hops, but there is more to aroma than what is suddenly special today.

Six years ago, Johann Pichlmaier, president of the Association of German Hop Growers, talked about the frustration farmers experience when the value of their hops is based on their bittering potential. "We don't like it when the discussion about hops is focused only on alpha acids," he said. Rereading that remark in an old notebook I realized it would also be a mistake to limit the discussion to a particular kind of aroma, because I also remembered something Italian winemaker Antonio Terni said in *The Accidental Connoisseur:* "I will only say that Americans like too much in the glass. There's always *too much* going on. Other than that, if we're living on Planet America, that's not necessarily the fault of Americans."[4] It would be silly to paint flavorful American beers with such a broad brush, but it is also true that new wave hops would not be as attractive if everybody enjoyed them.

Back at the conference table in Milwaukee, Ryder considered the most basic questions: How much more is there to learn? How long will it take? He shook his head from side to side.

"We sure ain't there yet."

He smiled. I tried to. That's not the answer you want to hear when you are writing a book.

About the Book

Synergy and perception. Neither is easily measured nor explained, but they are vital parts of any discussion of hops. When brewers, breeders, farmers, hop processors, and, of course, consumers focus on any particular aspect of the hop, the effects ripple through the production process.

This book could easily have started with the history of the plant or the chemistry involved, with a new season in the field, or with a brewer assembling a recipe.

Instead it begins with aroma, just like most current conversations about hops. The first chapter provides a primer on essential oils, the production of odor compounds, and how the human sensory system and brain turns those into aromas. There is no single formula. Brewers "want a checklist that matches oils and flavors," said Peter Darby of Wye Hops in England. "It's not that simple." As scientists learn how the sense of smell works and connects with what we call flavor, it becomes even more complicated.

The second and third chapters examine the plant's past and future. Were the book called *Romancing the Hop* it would include far more history. The plant has plenty of fascinating back stories, starting with the discussion in Chapter 2 about how it became an essential ingredient in beer. A complete history would amount to a volume more than twice this size; the Czech hop museum in Žatec fills 4,000 square meters and barely deals with anything beyond Bohemia. Histories of breeding, cultivation, and the trade could all be separate books. I would certainly buy one covering only landrace varieties, maybe one entirely about England's Golding and its many sisters. Instead, the chapter details how hops emerged as a vital ingredient and the varieties that gained particular fame before plant breeding dramatically changed hop growing. Discussions about new hop varieties and the future naturally direct attention to hop breeding, the subject of Chapter 3. There's much more to the lengthy process of getting a variety to market than creating "the flavor of the month."

Chapters 4 and 5 focus on the farm, growing hops, then harvesting and drying them. Brewers call cones, or even pellets, hops, but the plant itself mesmerized Charles Darwin, following the sun as it climbs, growing up to a foot in one day. Because vegetative growth and flowering depend on day length, the plant flourishes only at certain latitudes, and more recently scientists have been able to explain why precisely *where* a variety is grown changes the character of cones themselves. The farmers' work is not done until a crop is harvested and dried, and kilning is as important as any other stage of production in determining the brewing quality of a hop. Small breweries may not have the resources to send a representative to a hop growing region to pick particular lots of hops, but understanding the selection process is important. John Harris, who led the selection team at Full Sail Brewing for 20 years before leaving to start his own brewery in 2012, provides step-by-step directions for choosing the best hops.

Most traditional brewers use hops only in their "whole" form, either cones or pellets, although many now make an exception for CO_2 extract. Chapter 6, The Hop Store, includes a summary of all the forms available to brewers and provides vital information about and descriptions of 105 hop varieties.

The hop arrives in the brewery in Chapter 7, the first of three that look at the chemistry of the hop; extracting, calculating, measuring, and understanding bitterness; the results of different additions throughout the brewing process; and ways brewers may maximize the benefits of using hops. The eighth chapter deals specifically with dry hopping, both how brewers add hops post-fermentation and all the variables they consider. Chapter 9 includes what Boston Beer Company director of brewing David Grinnell calls, "practical, unsexy details"—measures brewers may take to assure quality, the benefits hops provide in sustaining beer quality, and the possible pitfalls.

In Chapter 10 brewers provide recipes that illustrate how they use hops. To explore the role of hops in various styles extensively would take another volume and is a reason those styles merit their own books. Instead, the recipes that follow illustrate how a few brewers include hops within the context of what we really care about—beer. These include beers hopped with particular enthusiasm, but those looking for information about India pale ale, the style that has focused new attention on hops in general and aroma in particular, should consider *IPA: Brewing Techniques, Recipes and the Evolution of India Pale Ale* by Mitch Steele.

In the late nineteenth century an uncredited English writer observed, "Fashion takes strange freaks, and it will be well for brewers to be prepared for all eventualities." The future of hops depends not only on the future of beer fashion, but also on the way brewers communicate with the hop industry, further scientific discoveries, and other factors. Brewers who produce a relatively small portion of the world's beer have made hops a bigger part of the beer conversation, but it could change again. There are no predictions about future fashion in the final chapter, but there are some thoughts from participants who will have a direct impact on "what's next?"

Notes

1. Processors usually package extract based on the amount of alpha acids in a drum. Most often, high alpha varieties, which tend to be better yielding, are extracted. So a field of Herkules, or Columbus in the United States, would produce more than one drum of extract. The point remains the same. An acre of hops strung seven meters (almost 23 feet) high occupies more than a million cubic feet. The majority of the brewers in the world are interested in only the tiniest portion of that.

2. D.R.J. Laws, "A View on Aroma Hops," 1976 Annual Report of the Department of Hop Research, Wye College (1977), 60-61.

3. F. Van Opstaele, G. De Rouck, J. De Clippeleer, G. Aerts, and L. Cooman, "Analytical and Sensory Assessment of Hoppy Aroma and Bitterness of Conventionally Hopped and Advance Hopped Pilsner Beers," *Institute of Brewing & Distilling* 116, no. 4 (2010), 445.

4. Lawrence Osborne, *Accidental Connoisseur* (New York: North Point Press, 2004), 19.

1

The Hop and Aroma

*The legend of BB1, and why you smell
tomahto plants and I smell tropical fruits*

In the spring of 1917 Ernest S. Salmon, a professor at Wye College, 60 miles east of London, placed a female hop in hill 1 of row BB of the Wye nursery. Salmon designated all his breeding material based on its position in the hop garden. He labeled the rows A, B, C, and so on; then AA, BB, CC. When he planted a wild Manitoban hop in hill 1 of row BB its name became BB1. BB1 matured early in the summer of 1918, flowering in July, forming large, coarse, somewhat pointed cones, many of them with undesirable leafy outgrowths. In the fall Salmon harvested the seeds the cones produced.

Salmon took charge of hop breeding at Wye in 1906, two years after the college began its program. An expert plant pathologist, he was already well known for research he conducted on powdery mildew but did not focus solely on breeding varieties resistant to that infection. In 1917 he presented a paper to the Institute of Brewing in London that revealed his main objective was to combine the high resin content of American hops, including some found growing wild, with the aroma of European hops. This plan would take hops in a new direction.

Although by early in the twentieth century American hop growers exported more than 10 million pounds annually, with 80 percent of those going to England, most brewers used them with reservations. American hops contained a higher percentage of alpha acids and thus had greater "keeping power" than English varieties. However, the opinion on those hops had not really changed since a harsh critique in *The Edinburgh Review* in 1862: "American hops may also be dismissed in a few words. Like American grapes, they derive a coarse, rank flavour and smell from the soil in which they grow, which no management, however careful, has hitherto succeeded in neutralising. There is little chance of their competing in our markets with European growth, except in season of scarcity and of unusually high prices."[1]

This recipe book, which brewing director John Keeling keeps on a shelf in his office at Fuller's Chiswick brewery, indicates that Fuller's used 5 percent Oregon hops to brew AK no. 70 in 1906.

Salmon had already cross-pollinated female American Cluster hops with European males, as well as European female hops with American males, when Professor W.T. Macoun, dominion horticulturist for Canada, sent him a cutting from a wild hop collected from beside a creek in the town of Morden in southern Manitoba. Macoun wrote, "Old residents in this town assure me that there has never been any introduction of cultivated hops in this district. The wild variety, growing so abundantly along the creek, was transplanted on the town lots, especially along the fences and back lanes, to cover unsightly places."[2]

BB1 did not take to its new environment, dying during the winter of 1918-1919, and two cuttings from BB1 also did not survive long in the nursery. BB1's seedlings would turn out much different.

Farmers and breeders propagate hops from cuttings because they do not breed true from seed; every seedling is genetically unique. Research by the Austrian monk Gregor Mendel in the mid-nineteenth century (which went unrecognized until about 1900) established that, contrary to Charles Darwin's theories, certain traits may occur in offspring without any blending of parental characteristics. His principles laid the foundation for plant breeding programs that created entirely new varieties. The seedlings of BB1 resulted from open pollination. They were bastards, not uncommon in hop breeding, although there's every chance the father was a Golding or Fuggle.

Salmon raised hundreds of BB1's offspring in a greenhouse beginning in 1919, planting some of them out in the Wye nursery in 1922, including one each in hills C9 and Q43. Because a previous plant in C9 had shown promise, he named the next one C9a. By 1925 that plant attracted attention for the richness of its cones, which, when opened and rubbed, became greasy or buttery to the touch. Salmon quickly propagated it, growing C9a on a larger scale at multiple locations, soon producing enough cones to be used in annual brewing trials. Analysts at the nearby East Malling Research Station, where additional trial plots were planted, used a formula based on the alpha acids and beta fraction to determine the preservative value of C9a, which Salmon would name Brewer's Gold. Both C9a and Q43 consistently recorded higher values than the richest imported American varieties.

Brewing trials produced mixed results. Following a test in a Kent brewery, head brewer C.W. Rudgard wrote: "For purposes of comparison a blend of Oregon and (Bohemian) Saaz hops was used. On comparing the finished ales it was found that in all ways the C9a was equal to the 'Oregon-Saaz' ale, and when considering the two ales from the aroma and hop flavor standpoint, there was a delicacy of bouquet in the C9a ale."[3] Compared against East Kent hops in a similar test, C9a also rated higher.

However, J.S. Ford of Wm. Younger & Company called C9a unsuitable for pale ale production, although he concluded that because of its preservative value it might be satisfactory in certain districts when used in small amounts or in a blend. Other brewers used descriptions that didn't really provide information about flavor, such as "Oregon," "American," and "American tang." One drew a distinction between "American" and "Manitoba" aromas and another time referred to a "pungent Manitoba aroma."

Salmon made Brewer's Gold available for commercial cultivation in 1934 and in 1938 released its sister, Q43, calling it Bullion. Farmers in the United States and Canada were quick to plant the varieties, which yielded alpha acids of between 8 and 10 percent in America, but the hops never received wide acceptance in England. It's not clear if brewers rejected them because they thought them truly objectionable or simply because they were different and bolder. One said that a general characteristic of the "American hop is that it is rather a stronger bitter than the European hop, and with a stronger scented flavor in many cases."[4]

More specific complaints referred to "catty" and "black currant" flavors. It would be decades before scientists would discover that a compound called 4-mercapto-4-methylpentan-2-one (otherwise referred to as 4MMP) is a main contributor to muscat grape/black currant character associated with American-bred hops such as Cascade and Simcoe. It has a low odor threshold and occurs naturally in grapes, wine, green tea, and grapefruit juice. Hops grown in the New World, including New Zealand and Australia as well as the United States, contain 4MMP and other compounds found only at trace levels in hops grown in England and on the European continent. Those other compounds are also associated with muscat/black currant character.

At one time, Salmon's cultivars accounted for about one-third of the world hop acreage. When he began at Wye College, hops contained 4 percent alpha acids on average and 6 percent at the most. Breeders have since released hops with more than 20 percent alpha acids, almost always using cultivars that lead back to Salmon. For most of the past century the focus remained on increasing alpha and replicating established aroma profiles. More recently, the definition of what constitutes a pleasant "hoppy" flavor broadened to include fruity and exotic flavors Salmon likely never envisioned. Yet almost every popular new variety—be it Citra, rich in passion fruit, or Mosaic, noteworthy for a distinct blueberry aroma—contains a bit of American wild hop, quite possibly BB1.

Hop Oils: Secrets Not Yet Revealed

David Ryder, vice president of brewing and research at MillerCoors, provides a concise list of seven positive attributes hops contribute in brewing:

- Bitterness
- Aroma
- Flavor (a combination of aroma and taste)
- Mouthfeel
- Foam and lacing
- Flavor stability
- They are anti-microbial, inhibiting the growth of organisms that damage the flavor and appearance of beer.

The genus *Humulus* includes three species, *Humulus lupulus, Humulus scandens,*[5] and *Humulus yunnanensis,* but the latter two do not produce resinous cones and are of no use in brewing. *Humulus* belongs to the family *Cannabaceae,* which also includes *Cannabis* (hemp and marijuana), and they are so similar that scientists have produced grafts between hops and hemp. End users likewise created a linguistic link between marijuana and hop-infused beers. For instance, drinkers sometimes describe beers dry hopped with particularly pungent varieties like Simcoe as "dank," a term long used to refer to potent, often still damp, marijuana.

When brewers talk about hops, they focus not on the plant but on the cone. The strobile develops from a female inflorescence (cluster of flowers). A zigzagging strig extends though the center of the strobile, and the strig bears a pair of bracts (outer leaves) and four bracteoles (inner petals) at each node. Lupulin glands develop at the base of the bracteoles. Lupulin itself is yellow, sticky, and aromatic. Cones range in size from less than an inch long to more than two inches and otherwise vary widely in appearance.

The lupulin glands contain hard and soft resins, hop oils, and polyphenols. The soft resins include alpha acids and beta acids, both of which contribute to bitterness. Until recently, scientists thought hard resins had no brewing value, but recent research indicates they can contribute a pleasant bitterness. Isomerized alpha acids, converted during wort boiling, are the primary source of bitterness (see Chapter 7). Hop oils are the key contributor to aroma and flavor, but the chemistry that creates aroma is not nearly as well understood as the chemistry involved in bitterness.

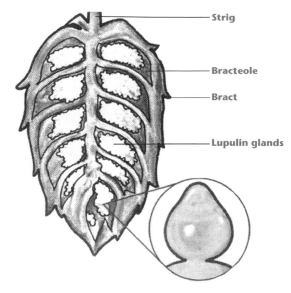

Strig

Bracteole

Bract

Lupulin glands

A dried hop cone includes 8 to 10 percent moisture, 40 to 50 percent cellulose, up to 15 percent protein, 8 to 10 percent ash, 2 percent pectins, up to 5 percent lipids and waxes, 2 to 5 percent polyphenols, 0 to 10 percent beta acids, 0 to 22 percent alpha acids, and 0.5 to 4 percent essential oils. Brewers focus on the last two, although not exclusively.

Odor Compounds Found in Hops and Hopped Beers	
2-methylbutyric acid	cheesy
3-methylbutyric acid (isovaleric acid)	cheesy
3-mercaptohexan-1-ol (3MH)	black currant, grapefruit
3-mercaptohexyl acetate (3MHA)	black currant, grapefruit
3-mercapto-4-methylpentan-1-ol (3M4MP)	grapefruit, rhubarb
4-mercapto-4-methylpentan-2-one (4MMP)	black currant
alpha-pinene	pine, herbal
beta-ionone	floral, berry
beta-pinene	piney, spicy
caryophylla-3,8-dien-(13)-dien-5-beta-ol	cedarwood
caryophyllene	woody
cis-3-hexenal	green, leafy

cis-rose oxide	fruity, herbal
citral	sweet citrus, lemon
citronellol	citrusy, fruity
ethyl-2-methylbutyrate	fruity
ethyl-2-methylpropanoate	pineapple
ethyl-3-methylbutonate	fruity
ethyl-4-methylpentanoate	fruity
eudesmol	spicy
farnesene	floral
geraniol	floral, sweet, rose
humulene	woody/piney
isobutyl isobutyrate	fruity
limonene	citric, orange
linalool	floral, orange
myrcene	green, resinous
nerol	rose, citrus
terpineol	woody

J. L. Hanin first used steam distillation to isolate hop oils in 1819, but it was not until the end of the nineteenth century that Alfred Chapman identified the key compounds myrcene, humulene, linalool, lynalyl-isonate, geraniol, and diterpene. He knew there were still more constituents to be discovered. "No one, for example, relying on the sense of smell alone could mistake Californian for Bavarian hops, or the latter for the produce of Kent," he wrote.[6] Because the composition of the compounds he identified did not explain those differences, he hypothesized that small quantities of other fragrant substances must be present. They simply could not be separated using technology available at the time.

Brewers certainly understood the importance of volatile oils. In 1788 in England, William Kerr patented a device that used a pipe to collect vapor leaving the kettle, then cooled the vapor before brewers separated the hop oil and water. They returned the oil to the boiling wort. Many similar inventions followed. One brewer specified: "One day before use the hops are soaked in a vessel of water until covered by the latter. There they remain at least 14 hours over a low fire, well covered, the heat in the tenth hour not to exceed 175° F. Then the mass is slowly boiled for 10 minutes, after which the liquid is emptied into the kettle with the first wort. The same process is repeated with a third wort."[7]

The introduction of gas chromatography in the 1950s revolutionized the analysis of hop oils and aroma. Researchers soon identified more than 400 compounds, a number that continues to climb. The contribution many of them make to finished beer is not at all clear because they occur at low levels, individually below perception thresholds. Together they often create distinctive odors through additive or synergistic effects.

Renewed interest in aroma touches every aspect of the hop industry, each step of the brewing process, and invariably includes discussions about oil as well. Breeders focus not only on the amount of essential oils but also the composition. Farmers plant different varieties and talk about varying harvest dates and kilning temperatures. Brewers work to extract more aroma from hops, or, more accurately, to pack more into their beers. The hop shortage of 2007 and 2008 made many more aware that hops are an agricultural product. "They want to learn everything they can twist and turn out of each handful of hops," Oregon grower Gayle Goschie said.

One lesson is simple. A brewer who visits a hop field two weeks before harvest may smell something much different in a cone than what is present by the time it is picked, and the compounds that form an aroma image in the brain change again during kilning and storage. Hop aroma in finished beer may be significantly different from that of a hop before it enters the kettle, and it changes as beer ages.

There is no single formula. "(Brewers) want a checklist that matches oils and flavors," said Peter Darby of Wye Hops in England. "It's not that simple."

Oregon hop merchant Indie Hops, founded only in 2009, pledged more than $1 million for what can broadly be described as aroma research at Oregon State University. "Our ultimate goal is to determine what is it in hop oil that drives flavor," said Thomas Shellhammer, who is in charge of the brewing science education and research programs at OSU. Shaun Townsend, who heads the Indie Hops aroma breeding program, would then use the information to develop cultivars with particular oil profiles. Additionally, if brewers can directly translate how the composition of oils in a hop variety, as determined by using gas chromatography, affects aroma and flavor, it would reduce the time and expense of brewing trials.

The research at OSU—which includes hop breeding, the influence of hop maturity on oil quality, dry hopping practices, and more—reflects an international focus on aroma that is different from what was being done even 20 years ago. In 1992 Gail Nickerson of OSU and Earl Van Engel of Blitz-Weinhard Brewing proposed establishing an Aroma Unit (AU) comparable to the International Bitterness Unit (IBU). The concept found little traction, but their suggestions moved the conversation about aroma forward, starting with the division of the 22 hop oil compounds that would compose the AU into three broad categories: oxidation products, floral-estery compounds, and citrus-piney compounds.

They intended that brewers would use their Hop Aroma Component Profiles along with the AU, much as they would use the alpha acid content of a particular variety to adjust hopping rates. "(Since the 1960s) scientists have tried to identify the compound responsible for hoppy character in beer without success. Hoppy aroma in beer is probably not attributable to a single component but rather to the synergistic effect of several compounds," they wrote in 1992.[8] That hasn't changed.

Tom Nielsen of Sierra Nevada Brewing explained how complicated it would be today to revitalize the concept to serve aroma-focused brewers. "You'd have to develop a citrus aroma unit, a floral aroma unit, a blueberry aroma unit … " he said, pausing as he considered the logistics. "You take three compounds, three aromas, different intensities relating to different human threshold levels. … You can't quantify that. It would take somebody doing a Ph.D." He later suggested it would be a particularly ambitious one.

The definition of "hoppy," never exactly clear, has broadened in the last 20 years. In Germany, for instance, the hop breeding program at Hüll Hop Research Center previously focused on maintaining the aroma quality of traditional hops such as Hallertau Mittelfrüh or finding new varieties with similar character. Now its mission includes a search for "unhoppy, fruity, exotic flavors derived from hops" and developing hops with those aroma traits. Anton Lutz, who is in charge of breeding at Hüll, made the first crosses for special aromas in 2006. The Society for Hop Research applied for plant variety rights for four new varieties in 2012, their aromas and flavors variously described as mandarin, passion fruit, honeydew melon, and, just as important, *intensive.*

In addition German-based Joh. Barth & Sohn recruited two beer sommeliers and a perfumist to compile the first volume of what will be three in *The Hop Aroma Compendium.* As well as systematically evaluating 48 varieties using traditional hop terms like "citrus" or "floral" the contributors also included descriptors such as "tonka beans" and "gooseberries."

Researchers in Hüll fully credit American craft brewers for driving interest in new aromas and flavors that result from compounds previously undiscovered or barely considered. Examples include:

- Flowery and floral (rose, geranium, clove)
- Citrus (lime, bergamot, pomelo)
- Fruity (tangerine, melon, mango, lychee, passion fruit, apple, banana, gooseberry, red and black currant)
- Vine bouquet (such as Sauvignon Blanc)
- Woody (pine, woody).

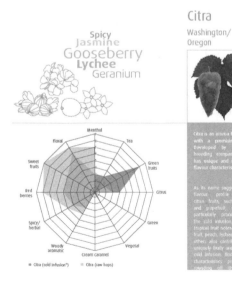

The Hop Aroma Compendium *provides two pages of information on each hop variety included. This is the first page describing Citra. The lighter shading on the spider chart represents the aromas found in "raw hops," while the darker one those perceived after hops were dissolved in water, thus simulating to some extent the change in aroma through dry hopping. Courtesy of Joh. Barth & Sohn.*

Hop oils, also known as essential oils, constitute up to 4 percent of the hop cone. They include 50 to 80 percent hydrocarbons, 20 to 50 percent oxygenated hydrocarbons, and less than 1 percent sulfur compounds. Hydrocarbons are highly volatile, not very soluble, and are perceptible in finished beer only when added very late in the boil or post-fermentation. The oxygenated compounds are more soluble and aromatic. Their aromas, or new ones resulting from the fermentation process, are more likely to show up in finished beer. Although sulfur compounds are a small portion of the oils, they have very low levels of threshold and can positively and negatively influence the aroma in finished beer.

Hop catalogs routinely provide information about the four most prominent oils in their variety profiles: myrcene, caryophyllene, humulene, and farnesene. The first is a monoterpene, meaning it consists of 10 carbon units, while the latter are sesquiterpenes (15 carbon units). Myrcene has a green, herbaceous, resinous aroma associated with fresh hops and not always considered desirable. It often constitutes 50 percent or more of the oils in American cultivars but is volatile, and most of its aroma is lost during boiling. Twenty years ago, standard advice for brewers who planned to dry hop with a myrcene-heavy variety was to let the hops sit warm for 24 hours to "mellow." Today many brewers and drinkers consider the pungent character it packs when fresh a positive attribute.

In their oxygenated form the sesquiterpenes are more likely than myrcene to survive into finished beer, their resulting aromas often described as "fine" or "noble." Caryophyllene and humulene are traditionally evaluated in tandem, with an H/C ratio of three to one considered a precursor for herbal and spicy aromas. Because farnesene content varies greatly, it is a good indicator of variety. It can constitute up to 20 percent of Saazer-type hops and is closely associated with "noble hop" aroma. Many other sesquiterpenes occur only in some varieties; for instance, Hersbrucker contains several not present in other cultivars.

As a hop ripens, many other monoterpenes form along with myrcene, their presence often measured in tenths of a percent—minuscule when compared to myrcene levels in some American cultivars. This is not a new discovery. Most compounds were components in the Aroma Unit but are receiving new attention because of citrus, fruity, floral, and woody aromas they produce. The Hüll Hop Research Center included data about many of them—including linalool, geraniol, nerol, citronellol, isobutyl isobutyrate, and limonene—as part of the profiles for its newest releases, and brewers may ask for similar information about all varieties in the future.

Research supported by Japanese brewer Sapporo examining how geraniol metabolism might add to the citrus flavor of beer reflects such interest. Researchers brewed two beers, using Citra hops in one and coriander seeds in the other because both are rich in geraniol and linalool. The finished Citra beer contained not only linalool and geraniol but also citronellol, which had been converted from geraniol during fermentation. The same transformation from geraniol into citronellol occurred during fermentation of the beer made with coriander. Taste panels perceived the beers as relatively similar. The concentration of geraniol and citronellol in both increased depending on the initial concentration of geraniol. The results suggested the importance of citronellol and an excess of linalool in the hop-derived citrus flavor of beer, but because there was little citronellol in raw hops the generation of citronellol depended on the geraniol metabolism by yeast.[9]

Scientists agree about the need for further studies focused on such interactions, referred to technically as biotransformations, between hop oil-derived compounds and yeast. Summarizing existing research completed as recently as 2011, Belgian scientists concluded most studies were limited to yeast and monoterpene alcohols, and few investigated the influence of yeast in brewery conditions. Those seem particularly necessary. The summary emphasized the need for further investigation into the transformation of oxygenated monoterpenes and sesquiterpenes as well as the evolution of hop glycosides during fermentation (p. 194).[10]

The importance of synergy often makes the contribution of individual compounds harder to measure. For instance, German researchers reported a mixture of caryophyllene and nerol had a flavor threshold of 170 parts per billion, compared to single thresholds of 210 parts and 1,200 parts per billion, respectively. The same was true of other mixtures, such as farnesene and linalool. The ratio of the blends also changed the threshold of perception.[11]

This might frustrate brewers who are looking for single markers. Nonetheless, Val Peacock, who has worked intimately with hops for most of the past 30 years, recently pointed out the danger of overestimating the value of any one hop component.

In 1981 Peacock and associates at Oregon State had suggested the importance of linalool to hop aroma. They developed a model to predict the amount of "floral hop aroma" likely in a beer based on the amount of linalool, geraniol, and geranyl esters in the essential oils. Scientists learned more about the compound, both its value as an indicator and its limitations, during the intervening years. Peacock noticed "because of the amount of attention given to linalool during recent decades, its perceived importance has been elevated far beyond its true relevance."

As a result, he concluded, "this distorts brewers' understanding of the nature of hop aroma in beer. Hops have more to contribute to the aroma of beer than just linalool, and except for one specific type of hop aroma, linalool is only a minor contributor to overall hop aroma."[12] Other researchers evaluating the hop character of Pilsner-type beers reported the impact of linalool depended on the form in which it was used. They discovered what they described as "high hoppy aroma intensity" was not always related to a high level of linalool. It was a reliable marker with conventional hopping, but not when brewers used advanced hop products.[13]

Whether in Edwardsville, Illinois, where Peacock keeps his consulting office, or Žatec in the Czech Republic, where Karel Krofta of the Hop Research Institute works, when chemists talk about hops in general or linalool specifically the word "complex" keeps coming up. "It is likely a carrier (of hop character), but not the only one," Krofta said. "Hop aroma has several hundred compounds. I don't believe only one can make that much difference." He quickly recapped many of the current discussions among hop chemists, talking about new research related to harvest dates, about the importance of kilning practices, about synergy and symbiosis. "It is not easy to study," he said. "The (chemical) pathways depend on many conditions along the way."

These are the sort of questions members of the hop group at MillerCoors continue to ask. "We have constant debates about linalool and geraniol," Ryder said.

"You can't say we'll add a little bit of this, a little bit of that," Pat Ting added, explaining that brewers make a mistake when they try to equate specific oils with specific odor compounds.

Less Is More and Other Aroma Secrets

More than two years after Marble Brewery opened in Albuquerque, New Mexico, director of brewing operations Ted Rice thought his only issue with *Marble IPA* was brewing enough of it. The beer accounted for nearly half of sales at the fast-growing brewery. Then one day early in 2011 he opened a bottle that was two months old and had been held in cold storage for purposes of quality assurance. He liked it better than a fresh sample in a side-by-side tasting. It still had the American hop "tang" British brewers detested 100 years before but more of the desirable fruit character he expected. This was the beer he wanted to sell fresh.

Rice and head brewer Daniel Jaramillo made changes one at a time, starting with the water. Albuquerque water is high in bicarbonates, and Marble installed a reverse osmosis water filter system before the brewery opened in 2008. They tried mixing RO water and city water in different proportions, everywhere from 100 percent RO to 25 percent. They changed the hopping schedule—for instance, making additions at 60 minutes, 30 minutes, and zero in one batch and first wort and zero in another. Then Rice considered a suggestion to reduce the quantity of hops in the recipe. While previously he had added more hops with the intention of increasing particular hop aromas and flavors, reversing that turned out to be the solution.

In February 2011 the recipe for *Marble IPA* included 1.65 pounds per barrel of hops added within the final 10 minutes (including knockout) and 1.8 pounds per barrel in dry hopping. In January 2012 the IPA included 0.8 pounds per barrel within the final 10 minutes and 1.2 pounds in dry hopping. Rice had the IPA he wanted. "I knew more was less, or less was more, but I wasn't really applying it," he said.

Aroma scientists established relatively recently why less may be more in the nose. A series of studies by Linda Buck and Richard Axel (who received the Nobel Prize for Physiology or Medicine in 2004) determined how the brain discriminates one odor from another. They discovered a family of 1,000 olfactory receptor genes that give rise to an equivalent number of olfactory receptor types. They later found that closer to 350 of the receptor types may be active, but even that number dwarfs the four types of receptors necessary for vision. About 1 percent of human genes are devoted to olfaction. Only the immune system is comparable, which is one reason smell is referred to as the "most enigmatic of our senses."

The olfactory receptors are buried in the two patches of yellowish mucous membrane called the olfactory epithelium, which are about seven centimeters up from each nostril. Humans have about 20 million receptors, covering the epithelium of both our right and left nostrils. The first stop a collection of molecules otherwise known as an odor makes on the way to the brain, and to being identified as a particular aroma, is in the receptors. Once activated, neurons transmit signals to the olfactory bulb of the brain, which relays those signals to the olfactory cortex. Olfactory information is sent from there to a number of other brain areas, including higher cortical areas thought to be involved in odor discrimination and deep limbic areas, which mediate the emotional and physiological effects of odors. Odor sensation becomes olfactory perception.

Buck and Axel determined odor receptors operate in combination to encode odor identities. Different odors are encoded by different combinations of odor receptors. Each odor receptor is part of the codes for many odors, and different odors have different receptor codes. Altering the molecular structure of an odor changes the receptor code and therefore the perceived odor. For the same reason, a change in the concentration of an odor may change how it is perceived. Higher concentrations involve additional odor receptors, again altering the odor response.[14]

When bottles of *Marble IPA* sat in storage, time apparently lowered the concentration of odorants. Nielsen's research at Sierra Nevada Brewing has shown several compounds that are common in beers with substantial amount of post-fermentation hopping, including beta myrcene, are among the first to be lost in or through crown liners.

Describing *Marble IPA* in *The World Atlas of Beer*, Stephen Beaumont wrote, "Where some breweries use American hops to attack the palate with citrus and pine, Marble cannily mixes flavors of overripe fruit with juicy and tangy hop notes to create a tantalizingly restrained ale that proves sometimes less really can be more."[15]

The sense of smell remains surrounded by multiple mysteries. Although the work of Buck and Axel revealed how odors are first perceived and how the brain translates them into discrete aromas, it does not account for how they are processed within the brain. Information gathered by the other senses is routed first through the interpretive reasoning centers of the left brain before heading to the emotional centers of the right side. In

Chinook

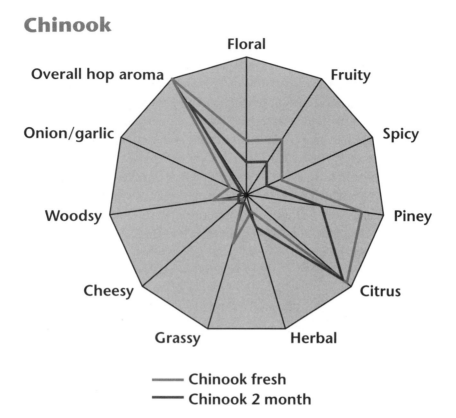

—— Chinook fresh
—— Chinook 2 month

Marble IPA is dry hopped with Cascade, Centennial, Simcoe, and Chinook, all varieties with compounds that may "disappear" relatively quickly. This spider graph, courtesy of Sierra Nevada Brewing, illustrates the difference between a beer freshly hopped with Chinook and one that is two months old.

contrast, whenever the olfactory bulb in the brain detects a smell, a chemical message is sent directly to the limbic system in the right hemisphere. The limbic system—including the hypothalamus, the hippocampus, and the amygdala—contains the keys to our emotions. Aroma may trigger memory, nostalgia, and mental pictures before the analytic left brain is first involved.

The most famous literary example of aroma-inspired memory would be French novelist Marcel Proust's account, in *In Search of Lost Time,* of childhood memories awakened by the scent of a madeleine dipped in tea. His 1913 novel was not the first to suggest a link between aroma and memory, but it since has been suggested he foreshadowed neuroscience. The cookie soaked in a decoction of lime flowers inspired powerful images, although little in the way of aroma descriptions: "But, when nothing subsists of an old past, after the destruction of things, alone, frailer but more enduring, more immaterial, more persistent, more faithful, smell and taste will remain for a long time, like souls, remembering, waiting, hoping, upon the ruins of all the rest, bearing without giving way, on their almost impalpable droplet, the immense edifice of memory."[16]

Hop Aroma Impact

Scientists disagree on how different aroma is from other senses in provoking memory or emotion. However, Japanese researchers supported by Sapporo Breweries determined the aromas of essential oils extracted from Saazer-type hops in fact exhibit a significant relaxing effect within the brain. They used headband sensors to measure the brainwaves of participants, including Sapporo employees who regularly drank beer as well as students who were unfamiliar with hop aromas. The results were similar for both groups, the rhythm of the brainwaves indicating subjects relaxed more when concentrations of hop aromas were higher. When subjects smelled particular essential oils, they relaxed more when smelling linalool or geraniol, but there were no changes when they smelled myrcene or humulene. The researchers had no explanation for the difference, but participants described the aroma of both myrcene and humulene as "green." A second study by the same group compared the effects of a variety of beers, most of them lagers. They found the aroma of Pilsner-type beer to be the most relaxing. More intense hop aroma showed a significant correlation with a relaxed feeling.[17]

Food scientists, like perfumists, have long worked to identify pure aroma chemicals that exhibit distinct character of the natural fruits, vegetables, spices, or other foods they were derived from. Oregon State University food scientist Robert McGorrin defined "character impact compound" as "the unique chemical substance that provides the principle sensory identity."[18] He also allowed that the overall impression might result from a synergistic blend of several compounds.

The blend may not seem nearly as complex. Andreas Keller and researchers at Rockefeller University determined subjects could identify only three (or rarely, four) components in a mixture. They invariably underestimated the number of odors in the mixture and perceived it no more complex than a single compound. In *The Scent of Desire* Rachel

Herz writes that the rose scent emanating from a flower bed contains between 1,200 and 1,500 different molecules, while just one molecule, phenethyl alcohol, imparts the scent of rose in many commercial hand lotions. Her research revealed that a majority of subjects were more likely to identify the synthetic versions as "the real thing," because those had become more familiar and "the prototypes for what we believe these fragrances *should* smell like."[19]

Tom Nielsen studied food chemistry at Rutgers University, his father was a flavor chemist, he interned at flavor house Robertet, and he might have ended up working for Pepsi or Campbell's Soup had he not been hired at Sierra Nevada. He views science and brewing each within the context of the other. His title, Technical Lead—Flavor | Raw Materials, reflects that. He's involved with flavor analysis, flavor research, flavor stability, flavor interactions, and all things malt, hops, yeast, water, and packaging materials. Nielsen likes to think the bar in his title marks a balance between the two areas. It also illustrates that the matrix that results in hop aroma is part of a larger complex enveloping beer aroma.

He has identified several key hop aroma compounds in beers, such as *cis*-rose oxide, prominent in Centennial hops, and caryophylla-3,8-dien-(13)-dien-5-beta-ol, which provides a distinctive cedarwood note to hops as they age. He is certainly interested in how particular compounds influence the aroma and flavor of Sierra Nevada beers, but like other hop and flavor scientists uses the word "synergy" often and points to the importance of understanding the physical interactions and biotransformations that occur in the presence of yeast.

He provided multiple examples to brewers during a presentation at the 2008 Craft Brewers Conference. For instance, he isolated four desirable fruity esters that are not present in unhopped wort or hops themselves but result from fermentation and aging of beer. In theory they arise from the breakdown of alpha and beta acid side chains and the subsequent cheesy-smelling, short-chain fatty acids. The flavor implications are significant but complicated, because a larger concentration of those short-chain fatty acids creates a greater potential for pleasant, fruity odors more easily detected (in other words, at lower thresholds) than the unpleasant cheesy acids.

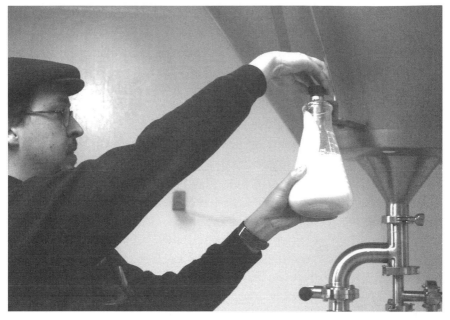

Tom Nielsen draws a sample from a fermentation tank at Sierra Nevada Brewing.

Myrcene Levels in Various Varieties

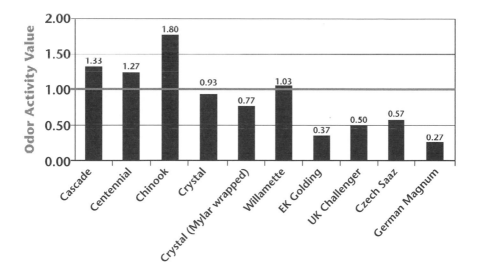

Nielsen also expanded on McGorrin's definition of character impact compounds, suggesting, "Highly hopped ales are very complex flavor systems with multiple character impact contributors." He explained how brewers may calculate "odor activity value" (OAV) just as food scientists do, dividing the concentration of a compound in beer by the threshold of that compound in beer.

Myrcene has a threshold of about 30 parts per billion in beer. Although it accounts for more than half of the oils in many American hops (for instance, up to 60 percent in Cascade), little will survive boiling and fermentation. In a controlled study comparing eight different hop varieties in a pale ale, Chinook had the highest myrcene OAV, 1.8, and Cascade the second highest at 1.3. Nielsen said that 400 parts per billion of myrcene, or more than 13 times the OAV, is a good target if aiming to create a strong, piney, resinous, highly hopped American beer. That is why it must be introduced post-fermentation, most often by dry hopping.

In 2003 Toru Kishimoto, at the brewing and research laboratory for Asahi Brewing in Japan, identified another major impact contributor specific to a family of varieties. "We had evaluated the hop aroma of many cultivars from the U.S., Europe, Australia, and New Zealand. At first mainly sensory evaluation," he wrote in an email. "Then I noticed the U.S., New Zealand, and Australian hops had a common fruity character in the aroma none of the European hops had." Using gas chromatography/olfactrometry analysis, he isolated 4-mercapto-4-methylpentan-2-one (4MMP). He does not know its source within the hop. "I still haven't investigated it. Given that the 4MMP exists with linalool or geraniol or esters in (the) fruity flesh of other plant species such as grapes or currants, 4MMP may exist with hop oils in the lupulin. I don't know the synthesis mechanism of the compound in the plant," he wrote.

Kishimoto tracked seven odorants that contributed to black currant aroma and identified two thiols beyond 4MMP: 3-mercaptohexan-1-ol and 3-mercaptohexyl acetate (3MH and 3MHA). They are also present in passion fruit and Sauvignon Blanc wines and are described as smelling of grapefruit and gooseberries. "These compounds contribute to the total sweet, fruity aroma and robustness of beer, though they don't contribute typical characters in beer directly," he wrote.

The amount of 4MMP in particular depends on variety and crop year. Kishimoto measured the highest levels in Summit, Simcoe, and Topaz, and it was not quite as prominent in Apollo and Cascade. Summit and Apollo exhibited a pungent, green onionlike, sulfur aroma in addition to black currant. Nielsen reported that 4MMP also occurs in Citra, Chinook, Bullion, and Cluster, and can be perceived at low thresholds, although perceptions vary widely. Descriptors range from "tropical" to "tomato plant." Further research confirmed what Kishimoto saw at the outset, that 4MMP was detected only in hops from the United States, Australia, and New Zealand.

Grown in the United States, cultivars such as Perle or Nugget contained 4MMP, but grown in Germany those same varieties had none.

He suggested this was probably because European farmers previously used copper sulfate (part of a "Bordeaux mixture" applied to both hop plants and grapevines) for protection against downy mildew, and it remained in the soil. He hypothesized that the undesirable, pungent, onionlike aroma common in cultivars such as Summit might be controlled by the use of copper sulfate.

He determined 3MH and 3MHA were different from 4MMP in several ways. Adding copper granules lowered the 4MMP content by 50 percent, but the 3MH and 3MHA contents remained unchanged, so, not surprisingly, 3MH appeared in both U.S. and European cultivars. The 4MMP content decreased 91 percent after boiling at 212° F (100° C) for 60 minutes. In contrast, 3MH content *increased* as the compound was heated to warmer temperatures. The results indicate 3MH is thermally formed during boiling from precursors that exist in hops. Levels of 3MH and 4MMP both increased during fermentation, peaking in the early stages.[20]

Further research in Japan has since isolated two more thiols known to otherwise contribute to wine flavors. Nelson Sauvin, a New Zealand hop with notable exotic fruit and white wine character, contains both 3-mercapto-4-methylpentan-1-ol (3M4MP), and 3-mercapto-4-methylpentyl acetate (3M4MPA). Researchers reported the compounds have a grapefruitlike and/or rhubarblike odor, similar to that of Sauvignon Blanc. The 3M4MP appeared at about two times its flavor threshold in beers, and although the 3M4MPA was below threshold it enhanced the aroma of 3M4MP by synergy.[21]

The Language of Aroma and Flavor

Nearly 100 years passed between the time philosopher Henry Finck proposed humans literally have a "second way of smelling" and University of Pennsylvania psychologist Paul Rozin established the role of retronasal smell in perception of flavor. In 1886 Finck suggested that smell was responsible for at least two-thirds of gastronomic enjoyment. In an essay titled "The Gastronomic Value of Odours," he began: "Amusing experiments may be made showing that without this sense (smell) it is commonly quite impossible to distinguish between different articles of food and drink. Blindfold a person and make him clasp his nose tightly, then put into his mouth successively small pieces of beef, mutton, veal, and pork, and it is safe to predict that he will not be able to tell one morsel from another. The same results will be obtained with chicken, turkey, and duck; with pieces of almond, walnut, hazelnut ..."[22]

Rozin conducted experiments in the 1980s that proved there are differences between orthonasal (breathing in) and retronasal (breathing out) perceptions of odors. Subjects who learned to identify smells by sniffing had difficulty recognizing them when they were introduced directly into the mouth.[23] Some of this is thought to be because of the way odors are first absorbed in the olfactory epithelium, differing based on the direction of the airflow across the epithelium. Retronasal smells activate parts of the brain associated with signals from the mouth, which helps to explain why we perceive flavor as occurring in the mouth even when the largest component is provided by what we smell.

That's one reason why a drinker might describe a beer as *smelling* bitter even though bitterness is a taste sensation perceived mostly on the tongue. Australian psychologist R.J. Stevenson discovered that after a novel odor is paired a few times with the sweet taste of sucrose, the odor is sensed as smelling sweet, because "sniffing the odor alone will evoke the most similar odor memory, that is, a flavor, which will include both the odor and the taste component."[24] Likewise, subjects described a novel odor as smelling sour after it was paired with citric acid. Stevenson and others attribute this to associative learning, also called conditioning.

In *Neurogastronomy: How the Brain Creates Flavor and Why It Matters*, Gordon M. Shepherd argues that retronasal is the dominant factor around which to build a field to study how the brain creates the sense of flavor (he suggests the phrase "human brain flavor system"), because retronasal smells are learned and unique to humans. Numerous other studies provide insights into the relationship between aroma and flavor, including that: 1) some cells are tuned specifically to odor; 2) some cells respond to both odor and taste stimuli, probably a step in creating the combined perception of flavor; 3) some cells respond preferentially to pleasant smells and others to unpleasant smells; and 4) preferences for smells can be learned and unlearned.[25]

This psychological interplay between aroma and taste that creates flavor has obvious implications for the overall hop impression of any beer, providing another explanation for the results of studies such as these:

- An experiment conducted by 35 members of the Rock Bottom Breweries group found no apparent relationship between measured bitterness and hop flavor or hop aroma, but a significant correlation between perceived bitterness and hop flavor or hop aroma. It appears that when drinkers smelled or tasted "more hops" they tasted additional bitterness even if the level of iso-alpha acids was the same (see p. 201).
- In a study in Belgium, scientists used hop oil fractions to create beers with different aroma profiles. Drinkers rated beers dosed with spicy hop essence higher in intensity of bitterness than beers without the essence even though they had equal bitterness units. In contrast, dosing beers with floral hop essence resulted in lower intensity scores. The spicy hop essence also enhanced mouthfeel.[26]

One challenge with beer in general, and, more specifically, hops, has been building a meaningful vocabulary. A beer flavor wheel, developed in the 1970s by the Master Brewers Association of America and the American Society of Brewing Chemists following the lead of Danish flavor chemist Morten Meilgaard, was one of the first such wheels. A wine aroma wheel came later, as did the Flavour Wheel for Maple Products, a South African brandy wheel, and many others. The Beer Wheel was not designed for consumers but to provide reference compounds that can be added to beer samples to represent the intended flavors. It continues to grow in size, and there is every possibility that the committees working on its redesign will settle on several subwheels.

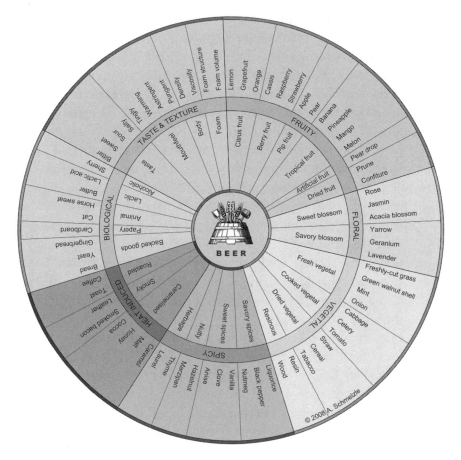

The Beer Aroma Wheel developed at Hochschule RheinMain University. Courtesy of Annette Schmelzle.

More recently, the Hochschule RheinMain University of Applied Science created a Beer Aroma Wheel (actually two wheels) with the goal of providing terms more suitable for communicating with consumers and focusing less on defects. Panelists who helped develop the terminology used aromas of fruits, spices, everyday materials, and other foodstuffs to describe their sensory impression. They often mentioned fruity, flowery descriptions—not exclusive to hops but often associated with hops—first. One version of the wheel (pictured here) was developed for consumers, a slightly more comprehensive one for industry use.[27]

Perhaps a hop-specific wheel will follow, but building a hop-centric vocabulary certainly presents a challenge. Contrast the adjectives in the *Hop*

Aroma Compendium that Joh. Barth & Sohn compiled with those used by Brewing Research Panel in England to evaluate products from Botanix (a subsidiary of Barth-Haas). In helping to compile the *Compendium* the sommeliers and a perfumist rated aroma using menthol, tea, green fruits, citrus, green, vegetal, cream caramel, woody aromatic, spicy/herbal, red berries, sweet fruits, and floral. In addition they added their own descriptors such as juniper, strawberry, and lemongrass (all for the Saphir hop). The BRP panel assessed beers in 22 categories, including tropical fruit and citrus fruit (a differentiation far easier for trained tasters than drinkers), pear/apple, herbal, grassy, and woody. They also include some that make sense mostly to trained panelists, such as linger, body, other sulfur, and DMS.[28]

Descriptions of bitterness relate more closely to what's perceived on the tongue, but, as Stevenson's research indicates, may become part of the language of both flavor and aroma.

They include bitter, of course, as well as astringent, chalky, medicinal, metallic, sharp, and aspirin.[29]

Why You Smell Tomahto and I Smell . . .

Recent evidence suggests that brain structures related to odor perception differ in size and cellular architecture between men and women. Women (on average) detect odors at lower concentrations, are more likely to rate smells as more intense and unpleasant, and are better able to identify them by name. Nielsen regularly notices the differences among members of the brewery's tasting panel. He said that when Sierra Nevada first began evaluating Citra, a hop rich in thiol compounds, men on the panel described tropical fruit flavors, while women called the same beer catty or said it reminded them of tomato plants.

Differences go beyond the sexes. Buck and Axel discovered many of the DNA sequences they identified are actually pseudogenes, meaning they are no longer functional. Instead of 1,000 active olfactory receptors, everyone has about 350 olfactory receptors. They aren't necessarily the same 350 receptors, providing a biological reason why two people will perceive a combination of odors, such as from a single hop variety, differently, or one of them might be altogether blind to a particular smell. "You end up with a barcode situation, whereby each individual has a slightly different barcode," said geneticist Doron Lancet of Israel's Weizmann Institute of Science.

Rockefeller University's Keller verified that genetic variation leads to differences in perception. He and his colleagues asked 500 people to rate 66 odors for intensity and pleasantness. The responses for any particular odor ranged from weak to intense and from wildly unpleasant to wonderful. "Everybody's olfactory world is a unique, private world," Keller said.[30]

The differences are not absolute and may change with training, an indication of the importance of what happens within the brain. For instance, although olfactory skills begin to deteriorate when most people are in their forties, many perfumers get better as they age. The most common estimate is that humans can differentiate between 10,000 and 40,000 odors. However, trained perfumers, whisky blenders, and chefs may be able to discriminate upward of 100,000 odors. One study examined brain activity in wine sommeliers and nonexperts as they drank wine. The sommeliers showed activity in areas associated with cognitive processing and in an area where taste and smell information is integrated. The nonexperts' activity was in the primary sensory areas and zones associated with emotional response. Research has shown practice making judgments about aroma leads to changes in brain function.[31]

Physical limitations certainly remain. Many people have specific anosmias: Although they have an otherwise normal sense of smell, they cannot detect a specific type of odor molecule. In Germany Thomas Hummel and associates tested 1,500 young adults with 20 different odors and discovered specific insensitivities to all but one—citralva, which has a citrus smell, and, coincidentally, is almost always considered a positive in hop aroma.

"The human genetics side turns the whole puzzle inside out and from gray to a full color spectrum at the same time," Nielsen said.

One-third of the population is blind to beta-ionone, a compound with a floral note particularly prominent in Saaz hops. Another third is extremely sensitive. Nielsen experiences that regularly. He began working at Sierra Nevada in 2004 after helping set up the Aromatrax gas chromatography-mass spectrometry-olfactometry unit that the brewery uses to identify odorants and link them to drinkers' perceptions. The invention of gas chromatography in 1955 allowed scientists to take a complex smell and pull it apart over time, creating a visual record of its volatile components. Individual odors that emerge from the GC are fed into the mass spectrometer, which came into use in the 1970s, to provide a definitive identification of the molecule.

Sierra Nevada panelists use a sniff port (the olfactometry unit) to smell an odor at the same time is it being separated, recording their impressions on a touchscreen pad. When Nielsen smells a specific caryophyllene alcohol compound, an oxidation product, at the sniff port, he perceives a very strong cedarwood note, an aroma many associate with "noble hops." He said about half of those he has tested are blind to it. In contrast, he is so sensitive to the compound he will sometimes notice the aroma lingering in a glass after he has finished a beer.

That's aroma impact.

Notes

1. Sydney Smith, et al., "Hops at Home and Abroad," *The Edinburgh Review* 116 (1862), 501.

2. E.S. Salmon, "Two New Hops: 'Brewers Favorite' and 'Brewers Gold,' " *Journal of the South-Eastern Agricultural College, Wye, Kent*, no. 34 (1934), 96.

3. E.S. Salmon, "Notes on Hops," *Journal of the South-Eastern Agricultural College, Wye, Kent*, no. 42 (1938), 54.

4. Ibid., 55.

5. In 2011 the World Checklist Programme in the United Kingdom decided to call the species previously referred to in all English-speaking literature as *H.japonicas* by its Chinese name, *H.scandens*, because of its Asiatic origin.

6. Alfred Chapman, *The Hop and Its Constituents* (London: The Brewing Trade Review, 1905), 65.

7. *One Hundred Years of Brewing* (Chicago and New York: H.S. Rich & Co., 1903, reprint, Arno Press, 1974), 90.

8. G.B. Nickerson and E.L. Van Engel, "Hop Aroma Profile and the Aroma Unit," *Journal of the American Society of Brewing Chemists* 50 (1992), 81.

9. T. Kiyoshi, Y. Itoga, K. Koie, T. Kosugi, M. Shimase, Y. Katayama, Y. Nakayama, and J. Watari, "The Contribution of Geraniol to the Citrus Flavor of Beer: Synergy of Geraniol and ß-Citronellol Under Coexistence with Excess Linalool," *Journal of the Institute of Brewing* 116, no. 3 (2010), 259.

10. T. Praet, F. Van Opstaele, B. Jaskula-Goiris, G. Aerts, and L. De Cooman, "Biotransformations of Hop-derived Aroma Compounds by *Saccharomyces cerevisiae* Upon Fermentation," *Cerevisia* 36 (2012), 126, 131.

11. C. Schönberger and T. Kostelecky, "125th Anniversary Review: The Role of Hops in Brewing," *Journal of the Institute of Brewing* 117, no. 3 (2011), 260.

12. Val Peacock, "The Value of Linalool in Modeling Hop Aroma in Beer," *Master Brewers Association of the Americas Technical Quarterly* 47, vol. 4 (2010), 29.

13. F. Van Opstaele, G. De Rouck, J. De Clippeleer, G. Aerts, and L. Cooman, "Analytical and Sensory Assessment of Hoppy Aroma and Bitterness of Conventionally Hopped and Advance Hopped Pilsner Beers," *Institute of Brewing & Distilling* 116, no. 4 (2010), 457.

14. Linda Buck, "Unraveling the Sense of Smell (Nobel lecture)," *Angewandte Chemie* (international edition) 44 (2005), 6136.

15. Tim Webb and Stephen Beaumont, *The World Beer Atlas* (New York: Sterling Epicure, 2012), 188.

16. Marcel Proust, trans. Lydia Davis, *In Search of Lost Time, Vol. 1: Swann's Way* (New York: Penguin Group, 2003), 47.

17. H. Kaneda, H. Kojima, and J. Watari. "Novel Psychological and Neurophysical Significance of Beer Aroma, Parts I and II," *Journal of the American Society of Brewing Chemists* 69, no. 2 (2011), 67, 77.

18. Robert McGorrin, "Character-impact Flavor Compounds," *Sensory-Directed Flavor Analysis* (Boca Raton, Fla.: CRC Press, 2007), 223.

19. Rachel Herz, *The Scent of Desire: Discovering Our Enigmatic Sense of Smell* (New York: Harper Perennial, 2008), 18-19.

20. Toru Kishimoto, "Hop-derived Odorants Contributing to the Aroma Characteristics of Beer," doctoral dissertation, Kyoto University, 2008, 68.

21. T. Kiyoshi, M. Degueil, S. Shinkaruk, C. Thibon, K. Maeda, K. Ito, B. Bennetau, D. Dubourdieu, and T. Tominaga, "Identification and Characteristics of New Volatile Thiols Derived From the Hop (*Humulus lupulus* L.) Cultivar Nelson Sauvin," *Journal of Agricultural and Food Chemistry* 57, no. 6 (2009), 2493.

22. Henry Fink, "The Gastronomic Value of Odours," *The Contemporary Review* 50 (November 1886), 680.

23. Avery Gilbert, *What the Nose Knows: The Science of Scent in Everyday Life* (New York: Crown Publishing, 2008), 93.

24. R.J. Stevenson, J. Prescott, and R. Boakes, "Confusing Tastes and Smells: How Odours Can Influence the Perception of Sweet and Sour Tastes," *Chemical Senses* 24 (1999), 631.

25. Gordon Shepherd, *Neurogastronomy: How the Brain Creates Flavor and Why It Matters* (New York: Columbia University Press, 2012), 115.

26. Van Opstaele, et al., 452.

27. Annette Schmelzle, "The Beer Aroma Wheel," *Brewing Science* 62 (2009), 30-31.

28. Thomas Shellhammer, ed., *Hop Flavor and Aroma: Proceedings of the 1st International Brewers Symposium*, (St. Paul, Minn.: Master Brewers Association of the Americas and American Society of Brewing Chemists, 2009), 80.

29. Ibid., 176.

30. Laura Spinney, "You Smell Flowers, I Smell Stale Urine," *Scientific American* 304, no. 2 (2011), 26.

31. Gilbert, 67-68.

2

A Plant With a Past

*How hops became a basic ingredient
in beer, and the varieties that emerged*

Workers widening a drainage ditch in the Graveney Marshes near Whitsable in Kent, England, in 1970 came across the remains of an Anglo-Saxon boat. Archeologists used radiocarbon dating to determine sailors unknown abandoned the 40-foot-long and 10-foot-wide vessel about 949 A.D. They found various plant samples, including considerable quantities of hops, that apparently had been part of the cargo. D. Gay Wilson of the University of Cambridge concluded that since cones made up most of the hop mass, the cargo was likely destined for brewing, even though the tenth century predates any other mention of hopped beers in England.

The discovery added one more date to a timeline that has few other documented entries before well into the second millennium. Wilson eventually concluded that much about the boat would remain a mystery, but along the way learned something else: "Beer is a popular subject, and the literature abounds in unsupported statements, misleading or inaccurate quotations, and inadequate references."[1]

A variety of colorful stories provide a deeper understanding of how important hops might have been in a particular beer culture, although they prove little about the evolution of hop usage. A few, such as the assertion that Jews were free from leprosy during captivity in Babylon because they drank beer made with hops,[2] appear unfounded and do not merit repeating. Many other accounts conflict with each other, and it is clear they will never be resolved.

The genus *Humulus* likely originated in Mongolia at least six million years ago. A European type diverged from that Asian group more than one million years ago; a North American group migrated from the Asian continent approximately 500,000 years later. Five botanical varieties of *lupulus* exist: *cordifolius* (found in Eastern Asia, Japan), *lupuldoides* (Eastern and north-central North America), *lupulus* (Europe, Asia, Africa; later introduced to North America), *neomexicanus* (Western North America), and *pubescens* (primarily Midwestern United States).

More recently hop scientists looking for genetically diverse germplasm analyzed wild hops found in the Caucasus, located between the Black Sea and the Caspian Sea. "Hops traveled quickly from China to Europe, but the (wild) hops of Caucasus are a little different," said Josef Patzak of the Czech Hop Research Institute, one of the authors of a study concluded in 2009. The researchers found the hop population of the Caucasus region genetically isolated from other European populations by barriers to migration or gene flow caused by climate or geographic changes.[3]

John Arnold suggested 100 years before, in *Origin and History of Beer and Brewing*, that native Caucasian tribes might have been the first to use hops in beer. Although observations about the Ossetians and Chewsures came from the nineteenth century, Arnold argued they were isolated from even the other tribes of the Caucasus and had been making beer "since time immemorial." Wilson discounted the conclusion, because by early in the Christian era knowledge about a universal practice such as brewing beer would have traveled efficiently via trade links that existed between the Roman world, Denmark and the Baltic, and Central Europe.

Descriptions about the importance of beer and hops in those tribal societies don't prove they were the first to use hops but do illustrate how hops become embedded in any beer culture. "In Western Europe the hop-plant and hopped beers have never penetrated so deeply into the soul and sentiment of the people, into folklore, folksongs, proverbs, sayings, legends, superstitions, and incantations, as has been the case with the peoples of the Northeast and East of the Continent, or at any rate not until much later," Arnold wrote. In Caucasia beer played an important part in sacrificial ceremonies and worship, and tribal members considered sacred drinking vessels as valuable as any of their possessions.

Hops Beyond the Brewing Kettle

Almost all hops grown worldwide end up in beer, and this book focuses on brewing with them. However, hop scientists continue to search for ways the plant may be used extensively elsewhere. For instance, the hop Teamaker, developed by the USDA in Oregon, has almost no alpha acids but substantial beta acids with antibacterial properties, which make hops valuable both in sugar processing and as animal feed (to inhibit certain types of undesirable bacteria in the digestive system).

Researchers at Oregon State University first discovered the cancer-fighting potential of the flavonoid xanthohumol in the 1990s, as well as its other antioxidant properties. It naturally appears in beer at relatively low levels, so that a drinker would have to consume about 30 liters of beer per day to benefit. However, the pharmaceutical industry continues to be interested in finding methods to extract xanthohumol from hops. More recently, a European commission approved hops for treating excitability, mood distur-bances, and sleep disturbances.

Otherwise, outside of beer, hops appear mostly as a novelty—in liqueurs, in candy, the shoots served in restaurants much like asparagus (or pickled and sold in hop museum gift shops), or the cones packed into hop pillows, which have been sold as a homeopathic cure for insomnia for more than a century.

Physicians and herbalists used hops, although not in quantities that required serious cultivation, long before breweries discovered their brewing qualities. In 1653 Nicholas Culpeper wrote the hop was "under the dominion of Mars. This, in physical operations, is to open obstructions of the liver and spleen, to cleanse the blood, to loosen the belly, to cleanse the reins from gravel, and provoke urine." He described other virtues at length, including its ability to cure the "French disease."[1]

[1] Nicholas Culpeper, *The English Physician* (Cornil, England: Peter Cole, 1652), 68.

The Chewsurians used two words for hops, *swia* and *pschala*, neither of which look like a derivative of anything found to the west. They added hops found in the wild to the fermentation tank after boiling wort for several days. The Ossetians also brewed with hops that grew wild, and at least by the nineteenth century added them during the boil. In a song said to date to ancient times, a maiden tells her youthful lover: "I shall gather from alders the clinging hop-vine, the wherewithal of beer for thee!" The Ossetian word for hopped beer (*k'umäl*) was etymologically connected to the word for hops (*chumälläg*) but not to the one for unhopped beer. They even claimed hopped beer as their own invention, although history indicates otherwise.[4]

Nineteenth century philologists carefully studied the various derivations of hop-related words, some sorting them into related groups, but ultimately they established little that settles when cultivation began or brewers first used them. What is certain is that Swedish botanist Carl Linné, commonly known as Linnaeus, gave *Humulus lupulus* its scientific name in 1753, likely taking *Humulus* from *humle* (also *humall* or *humli*), the Swedish word for hop, and *lupulus* from the Latin word for hop.[5]

Because the pollen from hops and hemp are identical, it is difficult to use archaeological evidence to distinguish between the cultivation of hops and the cultivation of hemp, leading to considerable confusion about where and when hops were grown. However, preserved fruitlets and bracteoles in waterlogged sites provide evidence of when *Humulus lupulus* first appeared in various regions. No concentrations of material that would signal cultivation have been found from the pre-Roman Iron Age or Roman periods, although Roman writers made many references to Germanic and Celtic beer. The first discoveries large enough to suggest cultivation of hops or their use in brewing, observed in Western Switzerland and France, date to between the sixth and ninth centuries. By the eleventh and twelfth centuries the number grows significantly; sites range north and west of Central Europe and are plentiful in the Netherlands, Northern Germany, and the Czech Republic.

Written evidence indicates hops were well known by the eighth century and were raised in monastery gardens. In 822 A.D. the Abbot Adalhard of Corbie issued a series of statutes that indicate the monastery used hops in brewing. A key passage states that a tithe of each malting was to be given to the porter of the monastery, who also used malt he made himself. The same rule applied to hops, and if the ingredients were not sufficient for *making beer* then he would take steps to find enough elsewhere. The hops were collected in the wild, and there is no mention of hop gardens. The statutes do not otherwise mention brewing, so there is no indication of when in the brewing process hops would have been added.[6]

Although the spread of hop cultivation implies brewers used hops, it was another 300 years before the writings of Abbess Hildegard of St. Rupertsberg confirm they boiled hops along with wort and appreciated their preservative value. In *Physica* (circa 1150-1160) she wrote, "If thou desirest to make a beer from oats and hops, boil it also with the addition of 'gruz' and several ash-leaves, as such a beer purges the stomach of the drinker and eases his chest"; also, "Its bitterness, though, when added to beverages, prevents in the latter putrefaction and gives to them a longer durability."[7]

Hildegard did not hint at how brewers discovered the importance of boiling hops, and that seems likely to forever remain a mystery. In *Beer: The Story of the Pint*, Martyn Cornell called this "the great unanswered question in the history of brewing," and suggested one theory. "Perhaps it was a dyer somewhere in Central Europe who boiled clothes with hop cones for an hour or more to dye some cloth deeply, and then accidentally tasted the cooled dye water, to be pleasantly surprised by its bitter flavor," he wrote, drawing on the fact that hop leaves and cones were used as dye. He supposed she might have decided to see if the result was the same in beer and further discovered the preservative power of hops.[8] Cornell acknowledged that there is no supporting evidence but put forward the suggestion to emphasize that boiling hops for an extended period represented a dramatic change in the brewing process and ultimately for beer. Today this would be called disruptive technology, and it would be no surprise that an "outside agent" was the protagonist. In fact, there's no proof that knowledge of the benefits of boiling hops originated at a single location and spread from there. Brewers operating in far-flung regions may have made the discovery independently.

Although the advantages of using hops became clear, it would take centuries rather than decades for hopped beers to become dominant. In *Beer in the Middle Ages and Renaissance*, Richard Unger described how brewers in north German towns first needed time to master making hopped beers before they began to export them, and that the pattern was then repeated by brewers in other regions. Additionally, drinkers had to acquire a taste for a drink that was more bitter than what they were used to.[9]

Judith M. Bennett explained in *Ale, Beer, and Brewsters in England* that the story of the slow acceptance of hops by English brewers and drinkers "is a story of urbanization, immigration, capitalization, and professionalization. It is also a story of masculinization, for beer brewing was seldom pursued by women."[10] The English made a clear distinction between ale and beer, and only the latter included hops. On the Continent, the protagonists were a little different—*gruit* ale and hopped beer.

In *Sacred and Herbal Healing Beers,* author Stephen Buhner suggested an alternative theory of why hops prevailed over herbs and spices that brewers may have been using since they had begun making beer. He discounted the well known advantages of using hops and occasionally blurred history, but his view has received enough attention to consider when tracking the course that ale—with or without the addition of *gruit*—and beer took beginning about the tenth century.

Buhner wrote: "It is important to keep in mind the properties of *gruit* ale: It is highly intoxicating—narcotic, aphrodisiacal, and psychotropic when consumed in sufficient quantity. *Gruit* ale stimulates the mind, creates euphoria, and enhances sexual drive. The hopped ale that took its place is quite different. Its effects are sedating and anaphrodisiacal. In other words, it puts the drinker to sleep and dulls sexual desire."

He suggested a direct connection to the Protestant Reformation and concluded the "historical record is clear that hops' supplantation of other herbs was primarily a reflection of Protestant irritation about 'drugs' and the Catholic Church, in concert with competing merchants trying to break a monopoly and so increase their profits. The motivations were religious and mercantile, reasons not so different from the ones used to illegalize marijuana in the United States in the twentieth century."[11]

Accounts of brewing in Mesopotamia and Egypt include descriptions of herbs and spices added to improve the keeping properties of beer or its flavor, and pharmacists preparing medicinal drugs often made them up in beer. Brewers were already using many of the ingredients that were part of any *gruit* mixture long before the government established control over *gruit*. *Gruit* itself added flavor and a certain amount of preservative power to beer, its composition was kept secret, and it varied from one region to another. Brewers made *gruit* ale, like any other ale in the Middle Ages, in a single vessel. They poured water and malt together and heated them along with other additives, conducted a mash, sometimes boiled and sometimes did not, and ran wort into wooden troughs or barrels to ferment.

The government Charlemagne created in the ninth century gave the emperor power over unexploited land, by extension the plants that grew on the land, and thus the key ingredients in *gruit*. Most prominently these included bog myrtle (also known as sweet gale), wild rosemary, and yarrow. In the Low Countries, bog myrtle (*Myrcia gale*) leaves were picked in the wild, dried, and crushed before use. Bog myrtle and wild rosemary did not often grow in the same regions and provided much the same "sharp taste," so it is likely only one would be included in a particular *gruit*. Different mixtures, used by both rural and monastic brewers, included other plant material and spices, such as ginger, anise and cumin in Germany. Laurel leaves, marjoram, mint, sage, acorns, caraway, wormwood, tree bark, and seemingly almost anything else that might be found growing made its way into *gruit* ale in different countries.

"These were the ingredients which the *fermentarius* (an official in charge of *gruit*) of medieval times, the apothecary of the medieval brewer, mingled and jumbled together according to his sweet will," Arnold wrote, " … the time which neglected no herb or drug, no matter whether harmless or poisonous, in an endeavor to lend some new property or savor to the brew."

He reported wild rosemary had a "stimulating effect" but nothing that suggested church or public officials considered it a problem.[12] Buhner wrote about eighteenth century accounts from Scandinavia that describe how intoxicating beers dosed with bog myrtle could be, but of course quantities used in the Middle Ages were kept secret and cannot be compared. However, tax records indicate the *gruit* addition to an ale was considerably smaller than the hop addition that replaced it. Unger acknowledged that "some contemporaries may have thought different," but stated there was no evidence beer made with bog myrtle was more intoxicating or that *gruit* had a narcotic effect.[13]

The emperor granted the *gruitrecht* (*gruit* rights) to both secular and religious authorities. Control over *gruit* afforded counts and bishops in Holland, the Low Countries, Westphalia, the Rhineland, and the lower Rhine region what amounted to a monopoly on what brewers using additives were obligated to use in beer. Otto I awarded the monastery at Gembloers the oldest known grant, in about 946 A.D. Counts or bishops usually controlled the grants, and by the twelfth century often sold or leased the *gruitrecht* to towns.

Typical for a *gruithuis,* the one at Dordrecht in Holland included brewing equipment. A *gruiter* (*fermentarius* in Latin) oversaw mixing the *gruit,* collected the tax, and often did some brewing himself. He required brewers to bring all the malt they planned to use to the *gruithuis,* so that a *gruitgeld* (tax) could be charged based on malt used rather than the amount of *gruit.* The *gruiter* supplied *gruit* wet from a cask in Dordrecht, which further allowed him to keep the ingredients secret. After hopped beer arrived in Dordrecht in 1322, beer brewed with *gruit* was called *ael,* the newcomer *hoppenbier.* When brewers began producing *hoppenbier,* many towns required them to pick up their hops at the *gruithuis,* and if brewers bought their hops elsewhere they were still required to pay a *hoppengeld.* The *hoppengeld* gradually replaced the *gruitgeld,* more quickly in towns where beers were brewed with hops and for export. The tax collected on *gruit* still amounted to eight times that on hops well into the fourteenth century in Leiden, where brewers focused on the local market. In contrast *gruit* was not even mentioned on the 1470 tax rolls of Gouda and Delft, which brewed primarily for export.

Quite often town officials paid little attention to the technology of brewing and simply levied a fee on each barrel of beer produced, leaving it to brewers to pick whatever herbs they wanted for their *gruit* or to use hops. A description of brewing in France would indicate why they chose the latter: "French brewers for a number of years were accustomed to making beer without hops. They substituted for it coriander seed, wormwood, and the bark of boxwood; but the bad quality of the beer disgusted customers and compelled brewers to use hops."[14]

Brewers in Bremen, Hamburg, Wismar, and other parts of Northern Germany began shipping hopped beer early in the thirteenth century, creating trade for both beer and hops as well as increasing demand for hopped beers. The cultivation of hops spread to Scandinavia, with the clergy and monarchy often promoting increased hop production. In 1442 Christopher of Bavaria, who was also king of Denmark, Norway, and Sweden, decreed that all farmers were to have "forty poles for growing hops."[15] The government required registered farmers to set aside a portion of their land for cultivating hops. By 1491 hops accounted for 14 percent of the value of Swedish exports.

Celebrating Hops

The population of the Belgian town of Poperinge in West Flanders no longer doubles in size following an influx of migrant workers during hop harvest, but every three years about 25,000 people fill the streets of the city center during a September weekend. The highlight of *Hoppefeesten*, a hop and beer festival, is a parade that snakes through Poperinge, with marching bands followed by riders on horses, followed by floats, with young and old dressed to depict hop pickers from an era past and sometimes even hops themselves (see photo, color plate 3).

History comes to life, as it does at annual harvest festivals in the Czech Republic, England, and other hop growing areas. Most of those regions also have startlingly lively hop museums. Hopmuseum Poperinge, for instance, is located in the former municipal scales, and on each of four floors there are interactive stations with quizzes written for both children and adults.

The European Union contributed more than €1 million toward a complex in Žatec that includes the Hop and Beer Temple (*Chrám Chmele a piva*), a labyrinth made from packed vintage hop sacks, a brewery-restaurant, and a 4,000-square-meter hop museum. A newly constructed, 40-meter-high "hop lighthouse" looks like a torch when lit up at night, hop trellis poles at the top shining as brightly as flames.

The outside of the museum shimmers metallic during the day, and the building glows at night. It occupies four floors of a former warehouse and hop packing hall, each filled with photographs, engravings, small tools, large vintage equipment, ephemera, and other items providing a history of hops that a book would need 1,000 pages to match.

In contrast, the HopfenMuseum outside of Tettnang is located on a working hop farm. A 4-kilometer walking trail connects the *Kronen-Brauerei* in the center of Tettnang with the museum. Every two years about 30 breweries, including Boston Beer Company and Anheuser-Busch, set up along the trail for a hop hiking day, a festival that attracts about 12,000 people.

Bernhard Locher, former chairman of the Tettnang Hop Growers' Association, founded the museum, and today his family cultivates about 35 acres of Tettnanger hops on the surrounding farm. A walkway extends off of one building into a field (color plate 6), allowing visitors to stroll out, eye level with the full-grown plants. They can experience modern-day harvesting almost as closely as those doing the work, contrasting that to a very different history on display in the museum.

Exhibits include the markers pickers received in payment and used like money, colorful wooden masks worn during hop carnivals, and pictures that depict the tradition of *Die Hopfensau* ("hop sow" or "hop pig").

It dates to a time when men would cut down individual hop bines and toss them to pickers, who were women. Pickers were paid only for full baskets, so whoever received the last bine also got the right to pool the contents of all the other baskets into her own. "Often it went to the pretty women," said Lucas Locher, Bernhard Locher's son.

Nearly every museum offers similarly special displays that leave an impression as strong as the giant cone at the Deutsches Hopfenmuseum in Wolnzach. That 5-meter-high cone could double as a drive-through coffee hut. It was built on a scale of one to one million, and inside, various buttons release a mist full of the odors of different hops grown in the Hallertau region.

Hop Museum
Žatec, Czech Republic
www.muzeum.chmelarstvi.cz
www.chchp.cz

Deutsches Hopfenmuseum
Wolnzach, Germany
www.hopfenmuseum.de

HopfenMuseum Tettnang
Tettnang-Siggenweiler, Germany
www.hopfenmuseum-tettnang.de

Hopmuseum Poperinge
Poperinge, Belgium
www.hopmuseum.be

Hop Farm Family Park
Kent, United Kingdom
www.thehopfarm.co.uk

American Hop Museum
Toppenish, WA
www.americanhopmuseum.org

In 1321 Count William III of Holland reacted to the popularity of hopped beers by outlawing their import from Hamburg, meanwhile permitting Dutch brewers to use hops for the first time. He lifted the ban two years later, instead setting restrictions on imports and instituting new taxes on them. By 1369, 457 breweries in Hamburg produced hopped beer and shipped much of it to the Netherlands. Eventually, Dutch brewers produced beer that competed well with the imports from Germany and began to export to other countries.

That included England in the 1370s, starting in the eastern and southern coastal towns, primarily for consumption by immigrants from the Low Countries. The terms "Dutch" or "alien" were used to refer to almost anybody from the Low Countries or Germany, and for more than 150 years after the first "Dutch" brewers began operating in London at the start of the fifteenth century they dominated the trade. Brewing with hops demanded not only a new skill but a different process that required hopped beer be boiled separately.

Because hopped beers did not use as much malt, brewers could make beer less expensively. In *A Perfite Platform of a Hoppe Garden*, author and hop farmer Reginald Scot claimed, "Whereas you cannot make above 8 or 9 gallons of indifferent ale out of one bushel of malt, you can draw 18 or 20 gallons of very good beer."[16] Simply put, beer brewers needed about half the malt as brewers of ale because with the addition of hops higher levels of alcohol were not necessary to improve a beer's longevity. That easily offset the cost of hops, additional equipment, and extra fuel needed to boil them and extract their preservative power. Consider this math from *Ale, Beer, and Brewsters in England* (with s. being shillings and d. pence):

"To achieve this higher yield, beer brewers did, of course, incur extra costs for hops and fuel. William Harrison's wife, Marion, spent 10s. on malt, 2d. on spices, 4s. on wood, and 20d. on hops; if we assume that half of her fuel costs went to boiling water initially (a process required in both brewing ale and beer) and half to seething the wort in hops (a process required only in beer brewing), brewing beer rather than ale cost her an additional 2s. for wood and 20d. for hops. For that additional 3s.8d. (about 25 percent of her total costs for materials), she more than doubled her brewing output, producing about 20 gallons for every bushel of malt. This was a nice saving for a domestic brewer such as Marion Harrison, but for a commercial brewery, these higher yields from beer brewing were even better news; they translated into higher profits."[17]

At the outset, brewers of beer sold their output for less. In 1418 for example, a tun of ale for the English army in France cost 30s. and a tun of beer 13s.6d. Over time the difference disappeared and, as on the Continent, translated into more profit for brewers who used hops.

As hopped beers spread across Europe, they pushed the invisible border between beer and wine consumption south. Mild weather in the south of Germany had favored wine production there, but soon beer was the everyday drink in Bavaria. In the thirteenth century well-to-do Flemish drank wine at meals. By the fifteenth century they preferred beer, and hopped beer was a sign of status. Wine sales reflected the shift. In the first four decades of the fifteenth century, wine exports from the French region of Bordeaux fell to a level about 15 percent lower than those of the first four decades of the previous century.

In Bavaria the Munich brewing regulations of 1447-1453 stated that only barley, water, and hops would be used in making beer. What's referred to today as the *Reinheitsgebot* repeated that decree in 1516 but was not the reason that hops triumphed over *gruit* as an essential ingredient in beer. It signaled a *fait accompli*. Sales of beer in Germany and the Low Countries grew faster in the sixteenth century than at any time before industrialization.

By then in England, more native drinkers had acquired a taste for hopped beer, but ale remained distinctively hop free. Some, though not all, English ale brewers used many of the same herbs and additives as brewers on the continent, but there was no system requiring them to acquire a *gruit* mixture from the government or an agent. In fact, Cornell makes a strong argument in *Amber, Gold, and Black* that no herbs at all were used in much medieval English ale, providing evidence that London and Norwich ale were herb free.[18] Additionally, in 1542, at a time when London still had more ale breweries than beer breweries but the beer outsold ale, writer and physician Andrew Boorde stated that only malt, water, and yeast belonged in ale.

In his self-help book *A Dyetary of Health,* Boorde also wrote, "I do drinke ... no manner of beere made with hopes," and "Ale for an Englishman is a natural drink ... beer ... is a natural drink for a Dutchman." He insinuated beer would give English drinkers the same fat faces and bellies as the Dutch aliens. Milton Barnes, a writer who has been characterized as a satirist, suggested Boorde's bias against hops was based not only on concerns about health. He reported that when Boorde was studying in Montpellier he got so drunk at the house of a "Duche man," presumably on hopped beer which the "Dutch" made and drank, that he threw up in his beard just before he fell into bed. Barnes wrote that when Boorde woke up in the morning, the smell under his nose was so bad he had to shave his beard off.[19]

Boorde was not alone, but as other anecdotal evidence suggests, he was part of a shrinking minority. By 1520 for instance, the town of Coventry in the Midlands had a population of 6,600 and 60 breweries producing hopped beer. Although 58 ale breweries operated in London in 1574, as compared to 32 beer breweries, it appears that Londoners drank four times as much beer. Nonetheless, during the next century John Taylor argued, "Beer is a Dutch boorish liquor, a thing not known in England, till of late days an alien to our nation till such times as hops and heresies came amongst us, it is a saucy intruder into this land."[20]

The distinction between ale and beer remains a part of English culture that is not always appreciated elsewhere, but it ceased to be specifically defined by the presence of hops. In 1703 the *Guide to Gentlemen and Farmers for Brewing the Finest Malt Liquors* noted the difference had become a matter of degree: "All good Ale is now made with some small mixture of Hops, tho' not in so great a Quantity as Strong Beer, design'd for keeping."[21]

'We Like the Hop That Grows on This Side of the Road'

During the fourteenth century Emperor Charles IV made exporting hop cuttings from Bohemian plants punishable by death, an indication farmers understood that not all varieties were alike and already knew how to divide the underground stems to propagate more plants. That particular variety, likely an ancestor of Saaz, would have been a native hop first found growing in the wild. "Seven hundred and fifty years ago somebody decided this was a great hop," said John Henning, a research plant geneticist with the U.S. Department of Agriculture Agricultural Research Service in Oregon. "Through the years growers chose the best growing hop. It could have been a mutant."

These hops continued to evolve because of both environmental factors and natural cross-pollination. Farmers would have chosen to propagate sister varieties that occurred naturally in their gardens because of yield, brewability, resistance to disease, and the other criteria breeders use today. In reality, as Val Peacock explained, decisions were usually pretty basic. "We like the hop that grows on this side of the road. We're not so happy with the hop that grows on that side of the road," he said.

Bavaria's agricultural school at Weihenstephan had 60 hop varieties in its collection by the beginning of the twentieth century. In 1901 John Percival described 20 different varieties of hops in England. The list did not include several named by other writers but did mention many that likely were variants of Golding. Hop geneticists call them landrace hops, implying they reflect the area where they grow and adapted over time to that region. When breeders began to use cross-pollination to create new varieties they usually started from these genotypes because they had the qualities brewers liked. All varieties being cultivated at the beginning of the twentieth century were in fact landraces, but the list compiled by Josef Patzak, et al., for a paper assessing the genetic diversity of wild hops provides a more workable number to consider. The authors point to Fuggle and Golding from England; Tettnanger, Spalter, and Hallertauer Mittelfrüh from Germany; and Saaz from what is now the Czech Republic.[22] Those from the Continent took their names from the regions where they grew, those in England the names of the farmer who, presumably, discovered them.

These include what others refer to as "noble hops," a term that came into use only in the 1980s and for which there is neither a set definition nor agreement about which varieties should be included. "The other side is, 'not noble' doesn't tell you anything," Jay Refling of the MillerCoors hop products group said. "Noble goes to perception. Noble says something you are doing is more classical."

In contrast, "noble" meant something specific to Boston Beer Company founder Jim Koch long before Samuel Adams brewed its *Noble Pils* with Saaz, Tettnanger, Spalter, Mittelfrüh, and Hersbrucker hops. He speaks in excited bursts while describing what sets them apart for him, personally. "Elegant, almost symphonically complex," he said. "The aromatics, clean bitterness. Floral, spice, citrus. Pine, spruce, eucalyptus. Aromas you don't get in any other hop. In fact, in no flower."

Identifying what makes them different is easier than tracing their history with any certainty. The original Saazer variety had, and still has, a red bine. It was called "Saazer Red" or "Auscha Red" to distinguish from the "Green Hop" that also grew in the area. In 1869 brewers paid 50 *gulden* per *zentner* for "Auscha Red," compared to 28 *gulden* per *zentner* for the "Green Hop."[23] More than a half century later, when hop breeding pioneer Karel Osvald began making clonal selections to improve the agronomic qualities of Saaz, he made it clear the hop would have evolved.

"Current hop cultures are a mix of vegetative posterity of various genetic origins. Issue of varieties is absolutely unclear, and we can talk neither of Czech old-Saaz red-bine hops nor of Semš hops. Semš hops originate from old-Auscha hops, which comes from old-Saaz hops," he wrote. "Today the cultures of hops are thoroughly mixed, even though there are no great differences between neighboring plants in a hop yard. Hops are plants that can adapt to growing conditions, since the soil and the climate lend them a certain character."[24]

In fact, DNA fingerprinting has since proved a genetic relationship so close between Saaz, Spalter, and Tettnanger hops that it almost appears they originated from the same plant. They display different morphological traits and brewing characteristics, thus in essence become different hops, because of the different environments where they are grown (p. 101).

A massive fire in the town of Žatec (Saaz in German) in 1768 destroyed historic documents that might provide more information about early cultivation, which apparently began in the first millennium. Few references to Bohemian hops exist before the fourteenth century, but from the time Charles IV actively promoted the product it flourished. In 1553 the Bohemian town of Plattan became the first outside of Spalt to receive its own hop seal, issued because dishonest merchants packaged inferior hops and sold them as "genuine Saaz."[25]

Although hop cultivation in the region of Spalt apparently dates to the late eighth century, the town rose to prominence in the trade after a monk from Bohemia introduced new methods of cultivation—and as genetic evidence now indicates, perhaps Saaz hops—in the fourteenth century. In 1511 the town banned the export of hop cuttings and shortly thereafter appointed certified hop inspectors at the market in Nuremberg. In 1538 Prince Bishop of Eichstätt granted Spalt the first German hop seal. During the next 300 years about 30 other hop-growing towns also acquired the right to issue hop seals.[26]

By the 1820s hops from Spalt demanded premium prices even in London. In providing a thorough survey of hops around the world in 1877, P.L. Simmonds strayed from what was mostly a report on quantities produced, and farming practices, to comment on the quality of Spalter hops. "The hops of Spalt are the best, and the German and French brewers who make beer which has to be kept long are obliged to employ Spalt hops with those of Saaz in Austria, which are the finest and most aromatic hops grown," he wrote. "The products are of a high reputation, and are the Chateau Lafitte, the Cose Vougent, and the Johannisberg, as it were, of hops of continental growths."[27]

The lineage of Mittelfrüh, on the other hand, is not so clear. Cultivation of hops in the Hallertau region continued on a small scale from the ninth century mention of hop gardens until the sixteenth century. In 1590 Kaspar Stauder planted imported hop cuttings from Bohemia at the Jesuit monastery of Biburg, near Abensberg. A priest himself, he later introduced hops to other farms operated by Jesuits, and perhaps these are the ones that became known as Hallertau Mittelfrüh or simply Hallertau. Even if they may have first come from Bohemia, they are genetically different today from the Saazer-type hops, with different brewing characteristics.

Refugees from Flanders established England's first modern hop gardens (and likely the first at all) relatively early in the sixteenth century, planting Flemish Red Bine hops. The color of bines differs according to variety, as do their stripes. Hallertau Mittelfrüh, for instance, has a green-red bine with a red stripe, while Spalt Spalter has a green bine with a green stripe. Red Bine was a landrace hop, obviously, surely different from those in Bohemia or Germany but likely sharing similar attributes. A study assessing the genetic diversity of wild hops concluded those found in North America vary widely, but European wild hops are not as variable and cluster closely with well known landrace hops. Grown in English soil, Red Bine did not produce desirable lupulin or much of it.

English farmers cultivated several other varieties by the beginning of the eighteenth century, the new ones a result of introducing more plants from the Continent or of natural cross-pollination (with native hops or imported plants). John Mortimer listed four specific hops in the

Whole Art of Husbandry—Wild Garlick Hop, Long Square Garlick Hop, Long White, and Oval Hop—and more generally two Whitebine varieties and a Greybine. Later in the century, a farmer near Maidstone spotted a Canterbury Whitebine "hill of extraordinary quality and productiveness, marked it, propagated it, and furnished his neighbours with cutting, from its produce."[28]

He gave the hop his own name, Golding, and by the beginning of the nineteenth century it had been distributed to all parts of England. Percival wrote, "It is clear that, although derived from the Canterbury Whitebine, the Golding hop was a specially selected sort, which had distinct characters of its own."[29] Additionally, in 1799 a Kent farmer described Canterbury Whitebines, Farnham Whitebines, and Golding hops as distinct varieties.[30]

Until recently Percival's account about the origin of the Fuggle hop—that it was grown from a seed shaken from a hop-picking basket in 1861 and introduced by a Richard Fuggle in 1875—made for a colorful story about England's other defining hop variety. Research reported in an article in *Brewery History* magazine in 2009 calls every part of the story into question, not altogether surprising with history and hops.[31]

Fuggle's place in history remains unquestioned. It became the dominant hop variety in England, accounting for 78 percent of acreage in 1949 before verticillium wilt nearly wiped it out. Oregon farmers grew primarily Fuggle at the time they produced half the hops in the United States. In Slovenia it became Styrian Golding, and some think the hop called U.S. Tettnanger is a Fuggle, which would make it a great-grandmother of Citra. Fuggle is also a parent and grandparent of Cascade, a grandparent of Willamette, and it may have produced the seedling that grew into the first dwarf hop at Wye College. Recent research indicates Osvald's Saaz clone 126 could have originated from Fuggle, although the clone's morphological traits and aroma components are much like Saaz and unlike Fuggle. It clearly adapts as it must, making it a plant with a future as well as a past.

Notes

1. D. Gay Wilson, "Plant Remains From the Graveney Boat and the Early History of *Humulus lupulus L.* in Europe," *New Phytol* 75 (1975), 639.

2. John Bickerdyke made this statement in *The Curiosities of Ale & Beer* (p. 26), but Wilson carefully explains why the connections Bickerdyke, and others, made do not hold up to scrutiny.

3. J. Patzak, V. Nesvadba, A. Henychova, and K. Krofta, "Assessment of the Genetic Diversity of Wild Hops (*Humulus lupulus* L.) in Europe Using Chemical and Molecular Markers," *Biochemical Systematics and Econology* 38 (2010), 145.

4. John Arnold, *Origin and History of Beer and Brewing From Prehistoric Times to the Beginning of Brewing Science and Technology* (Chicago: Alumni Association of the Wahl-Henius Institute of Fermentology, 1911, reprint, *BeerBooks.com*, 2005), 145.

5. In his book *Oranges* (New York: Farrar Straus and Giroux, 1967, 64) John McPhee illustrates how easily confusion may arise related to botanical names. The ancient Greeks called the citron tree *kedromelon*, or "cedar apple," when it arrived in the Mediterranean basin, because it resembled the cedars of Lebanon. The Romans turned that into *malum citreum* and applied the term *citreum* to all the various fruits of citrus trees. When Linnaeus made citrus the official name for the genus he grouped lemons, limes, citrons, oranges, and similar fruits under a name that means cedar.

6. Wilson, 644.

7. Arnold, 230.

8. Martyn Cornell, *Beer: The Story of the Pint* (London: Headline Book Publishing, 2003), 59-60.

9. Richard Unger, *Beer in the Middle Ages and Renaissance* (Philadelphia: University of Pennsylvania Press, 2004), 55.

10. Judith Bennett, *Ale, Beer, and Brewsters in England* (New York: Oxford University Press, 1996), 78.

11. Stephen Buhner, *Sacred and Herbal Healing Beers* (Boulder, Colo.: Brewers Publications, 1998), 173.

12. Arnold, 240.

13. Unger, 31.

14. P.L. Simmonds, *Hops: Their Cultivation, Commerce, and Uses in Various Countries* (London: E. & F.N. Spon., 1877), 94.

15. Ian Hornsey, *A History of Beer and Brewing* (Cambridge, England: Royal Society of Chemistry, 2003), 307.

16. Bennett, 85.

17. Ibid., 86.

18. Martyn Cornell, *Amber, Gold, and Black: The History of Britain's Great Beers* (London: The History Press, 2010), 173.

19. Cornell, *Beer: The Story of the Pint*, 70-71.

20. Bennett, 80.

21. Cornell, *Beer: The Story of the Pint*, 86.

22. Patzak, et al., 136.

23. H.J. Barth, C. Klinke, and C. Schmidt, *The Hop Atlas: The History and Geography of the Cultivated Plant* (Nuremberg, Germany: Joh. Barth & Sohn, 1994), 214.

24. From "100-year Birth Anniversary of Doc. dr. ing. Karel Osvald (*sic*)." Retrieved 23 August 2012 from *www.beer.cz/chmelar/international/a-stolet.html.*

25. Barth, 202.

26. Barth, 116.

27. Simmonds, 86.

28. William Marshall, *The Rural Economy of the Southern Counties* (London: G. Nichol, J. Robinson, and J. Debrett, 1798), 183.

29. John Percival, "The Hops and Its English Varieties," *Journal of the Royal Agricultural Society of England* 62 (1901), 88.

30. There is no way to determine if Golding has only wild English hops in its pedigree or a mixture of wild and Flemish Red Bine. Wye Hops has no authenticated examples of either the Flemish Red Bine or any wild English hops from the 1700s on which to base a comparative DNA analysis. The accession in the collection considered closest to the Flemish Red Bine is called Tolhurst. Unlike Golding, it contains farnesene, which might suggest an origin similar to Saaz or Hersbrucker.

31. Kim Cook, "Who Produced Fuggle's Hops?" *Brewery History* 130 (2009), 3-17.

3

A Plant
With a Future

*Aroma is in fashion, but hop breeders
still abide by the rules of agronomics*

Every pole in the test fields at the German Hop Research Center Hüll has a number, so Anton Lutz, who runs the breeding program, can quickly identify the grown plants he refers to as "my seedlings." In a month he would be in the field making evaluations, book in hand, so he could easily check on a plant's mother and father were he to come across a surprise.

This August day he walked up one row and down another. German brewers call him *der Hopfenflüsterer*, the hop whisperer.[1] "You look at his eyes and you see there is something always going on behind them," said David Grinnell, vice president of brewing at Boston Beer Company, who sits on the advisory board at Hüll.

Lutz grabbed a plant. "Only one shoot," he said, shaking his head. This one wouldn't be back in the field next year. He moved from one bine to the next.

"Climbing very bad."

"Died."

"Climbing bad."

"Too toppy."

"This seedling looks good. Maybe we harvest her. (Cones) come from bottom to top."

"A little downy mildew."

Not once did he reach for a cone and open it to check the aroma. Gene Probasco, who breeds hops in the Yakima Valley, could have been walking beside him. "The first real evaluation is visual. (A plant) has to look a certain way or it doesn't get a second look," he said. "I like to know what family I'm looking at. I have certain expectations for families."

Hop breeders appreciate, even share, the excitement about new special aromas but understand if farmers can't afford to grow a plant that has no future. They focus first on agronomics—on finding plants that yield well, are less susceptible to diseases and attack by insects, and can be easily harvested and stored. They use DNA markers and genotype fingerprinting and often work in modern laboratories, where they can instantly analyze a plant in a frightening number of ways, such as they do at the Czech Hop Research Institute in Žatec or the German facilities at Hüll.

Peter Darby had never seen a hop plant before he went to work at Wye College in 1981, examining the single gene attributes of hops. He previously had studied the inheritance traits of disease-resistant peas and leaf spot disease. Now he takes vacation in the weeks before he must decide which plants being trialed at China Farm in Kent will survive another year, and which he will reject. Otherwise, he would be in the field every day, forming opinions about plants before he really should. It's not complicated, he explained, using a story about his mentor and pea plants. "He had these canes and would load them up on his arm," Darby said, extending his left arm to illustrate. "When he'd come to a plant he wanted to keep, he'd plant a cane." Obviously, not many went on to the next round.

Advances in molecular biology have changed breeding, but there's still something very basic at the core. Darby makes his crosses much as E.S. Salmon did more than 100 years ago. He collects pollen from a male hop, adds it to the female's pollen sleeve, and closes the bag. In the fall he collects the seeds and the following spring plants seedlings in the greenhouse.

"Absolutely the same," he said. "Choosing the mother and father; all the creativity is in that stage." Looking at plants still full of the promise they held a few months before, he said, "When the seedlings go into the field in May, at this point a great hop is a great hop. But it's not identified. We haven't found a way to speed that up."

Are You My Mother?

The U.S. Department of Agriculture released the Chinook hop in 1985, a seedling selected from a cross 11 years before. Chuck Zimmerman crossed a female English hop and a male, labeled USDA 63012M, that was part of the breeding program at Prosser Station, Washington. The USDA acquired the female, Petham Golding, in 1968 from Wye College. 63012M had been selected from a cross between Brewer's Gold and a wild hop collected in Utah.

Recent DNA tests in Slovenia have confirmed that Golding is the mother of Chinook. However, when Ray Neve of Wye College visited Corvallis in 1977 for the International Hop Growers Convention Scientific Commission meetings, he was shown the plant reputed to be the Petham Golding mother plant. He didn't think it had the right cone shape and arranged for cuttings to be sent to Wye. Tests on the regrowth showed it was infected with hop mosaic virus. Had it been a true Golding, that would have killed it quickly. Later, oil analyses of the cones confirmed that the plant in the Corvallis collection thought to be the mother of Chinook was not a Golding. What has happened to the actual Golding plant that was used to create Chinook is a mystery.

During the nineteenth century a hop grower could expect a full crop only every 10 to 20 years. As a result an excess of acreage remained under cultivation, creating massive oversupply in bountiful years, shortages when crops failed, and wild price swings.[2] Attempts shortly before the turn of the century, in various hop growing centers, to apply newly established principles of plant breeding to hops did not last long. Wye College began its research program in 1904 and hired Salmon shortly thereafter, and it was his work that most influenced those who followed.

Dozens of public and private programs now operate around the world. Some of them, such as in New Zealand and Australia, recently released varieties with particularly unique aromas, but most research is related to combating new or old diseases, improving yield, making low-trellis systems viable, or other advances that serve growers. For instance, in South Africa, Mexico, and Colombia breeders work to develop varieties better suited for shorter day lengths. (The hop is a short-day plant, meaning it does not flower well in locations closer to the equator.)

Czechoslovakia resisted using cross-pollination longer than any other hop growing center, relying instead on clonal selections to improve its Saazer-type plants. Breeding pioneer Karel Osvald, who began the Czech program in 1925, feared attempts to increase yield by hybridization would result in loss of the aroma that characterizes Saaz hops. Maintaining the same aroma was the top priority, just as in programs almost everywhere else.

Osvald, who died in 1948, studied cross-breeding thoroughly and wrote in length about it, but the Czechs did not make crosses until the 1960s, and it was 1994 before they registered their first bred variety. Saaz accounts for about 83 percent of the Czech Republic's production, although acreage of new varieties such as Harmonie, Premient, Sládek, and Agnus recently increased. The Hop Research Institute released the first nonclonal Saazer type, Saaz Late, only in 2010. Scientists began working on the variety in the 1990s, using male genotypes that, like the mother, have their origin in Saazer, so its resins and essential oils are very similar to the original. However, its alpha acid content will not vary as wildly as in Saaz itself, which seems to be affected particularly by hot summers, and its yield is better. In addition to conducting seven years of comparisons with Osvald clones, the institute sent samples to 15 Czech breweries for brewing trials. "You have to rely on the experience of the breeder," hop chemist Karel Krofta said, explaining the process. "If the breeding period is too short, there is a danger some bad properties can be revealed in the course of cultivation."

The timeline that begins in Germany with cross-pollination and ends with an application for plant variety rights illustrates why hop breeders must be thinking ahead 10 years or more. "The big problem for breeders (worldwide) is aroma types," Lutz said. "Each brewer has his own specific idea. He wants to test it in his own brewery, to use his own water. You have to have a lot of aroma types ready, but you must wait for brewers to come to you with their ideas. Then you can tell them, 'I have it.' "

The Birth of a New Variety

At German Hop Research Center, Hüll

Year 1: 75 to 100 crosses.

Year 2: 100,000 plants are grown from seedlings in a greenhouse and assessed for their resistance to powdery mildew and downy mildew. Most will fail.

Years 3-5: 4,000 female and 400 male plants assessed in the breeding yard. Assessed again for resistance, and also for wilt, stature, cone quality, yield, and aroma quality.

Years 6-9: 20 to 30 advanced lines remain. Ongoing assessment of all traits, plus storage stability.

Years 10-12: Highly advanced selections, field trials at different locations, and brewing trials.

Years 13-14: Apply for plant variety rights.

At USDA, Corvallis

Year 0: Make crosses.

Year 1: Seedlings are grown in the greenhouse and selected for powdery mildew resistance.

Years 2-4. Plants are field assessed, evaluated and harvested, chemically analyzed. The Oregon Hop Commission and the Hop Research Council decide which look promising. The HRC was established in 1979 by brewers, dealers, and growers to fund and direct U.S. hop research. Boston Beer Company and Sierra Nevada Brewing became members relatively recently, and the Brewers Association joined the HRC in January 2012.

Years 5-8: Selections grown on multi-hill plots. Plots harvested for accurate yield date. Evaluation continues, complete data collected. Samples sent to breweries for pilot brews. Breweries select favorites.

Years 9-???: Selections grown in commercial farm plot (15-30 hills, or one-acre plots). Tested at multiple breweries. Brewers accept or reject the hop.

Lutz first made crosses for floral, fruity, and otherwise exotic aromas in 2006, and in 2012 the Society for Hop Research applied for plant variety rights for four new varieties. That's quicker than most varieties go from cross to field. (Brewers who tasted samples of beers made with two of the new hops at Brau Beviale in Germany in 2011 and the Craft Brewers Conference in the United States in 2012 showed immediate interest in both Mandarina Bavaria and Polaris.)

Scientists have many more tools available to them today than when the Germans established the center at Hüll in 1926 and the U.S. Department of Agriculture program began in Oregon in 1930. Molecular markers, for instance, make it easier to map for alpha acid content or resistance. However, the genes influencing many attributes interact in complex ways, so even when selecting for a few specific traits researchers must monitor nontarget features to guarantee that unwanted changes do not occur.

The institute at Hüll focused at the outset on developing disease-resistant plants with aromas similar to traditional varieties, identifying both cultivars and wild hops with limited resistance, crossing them, and then crossing them again and again to build up their resistance level. When verticillium wilt later threatened the future of Hallertau Mittelfrüh, their attention turned to wilt. The programs at Wye and Hüll benefited each other. Downy mildew-resistant genotypes from Hüll made it possible to introduce such resistance into the Wye program very rapidly, and Wye's Northern Brewer aided German breeding for verticillium wilt resistance. After farmers and brewers demanded high alpha hops, researchers at Hüll used cultivars from Wye, Yugoslavia, and the USDA along with its own breeding material to create Herkules, released in 2005 and noteworthy for both its alpha acids and yield.

The German government provides €1.8 million annually to support the Hop Research Center. The objectives at Hüll, constantly updated, serve as a reminder that although hops with special aromas dominated the conversations at the outset of 2012, research related directly to the plant entails much more. The goals included:

- Breeding fine aroma types. Research continues to focus on traditional Hallertau aroma. Additionally, in 2011 the center entered into a collaboration with Tettnang hop growers to develop a variety with an aroma profile similar to Tettnanger but slightly higher alpha acid content, higher disease resistance, and better yield.

- Developing varieties with special features, including not only special aromas but also hops with higher levels of beta acids, xanthohumol, anti-oxidative substances, and other health-promoting compounds.
- Developing dwarf hops with yields and other qualities suitable for growing on low-trellis systems. Because many German farms are small and family run, the economics don't make sense in Germany today but could in the future.
- Breeding for resistance to powdery mildew. Between 2001 and 2007 Hüll screened more than 15,000 wild hops collected world-wide for resistance to powdery mildew.
- Utilization of molecular markers to select for powdery mildew resistance and to differentiate male and female hops.
- A joint project with the Slovenian Hop Research Institute to detect verticillium wilt more quickly.
- Monitoring for hop stunt viroid and virus infections, not currently a problem in Germany but developing in other hop growing regions.

In 2007 Britain's government yanked funding for the program at Wye, which since 1948 had been an equal partnership between industry and the government. "We would have lost 100 years of experience," said hop farmer Tony Redsell, who spearheaded the establishment of Wye Hops Limited, headed by Darby. "He'd have been put on raspberries (at nearby East Malling Research)."

Wye Hops occupies about eight acres in the midst of Redsell's China Farm, one of his properties near Canterbury, and Redsell's employees do much of the field work, which the National Hop Association then pays for. Standing alongside Wye experimental hops, Darby pointed to some of Redsell's across a path. "On this side consistency," he said. Then he gestured to the test field. "On the other, diversity."

The collection at Wye contains not only classic cultivars such as Old Golding from 1790 and Fuggle from 1875, but wild plants from Japan, Europe, and North America, breeding lines from programs throughout the world, commercial varieties from public programs, and all the working parental material from the Wye program since Salmon's first crosses. The germplasm includes genetically distinct sources for resistance to downy mildew and wilt diseases. It also encompasses many novel characteristics and genotypes with chemical compositions well outside those found in commercial varieties. Libraries at Corvallis, Oregon, Hüll, and elsewhere differ but hold equally wide ranges of material.

Wye released both the world's first dwarf variety, First Gold, and the first aphid-resistant variety, Boadicea. British farmers began growing low-trellis hops in 1996, and now dwarf varieties occupy about 25 percent of acreage. Darby is working with the Czechs to develop more dwarf or low-trellis varieties. Additional projects in the Yakima Valley and at Hüll imply these alternatives have a future. Darby said that it appears that such plants grew at Wye throughout the twentieth century, although the first dwarf wasn't recognized until 1977. Breeding records from 1911 describe an unusual seedling of Fuggle that had "laterals of medium length, very closely placed," adding that it was "very fruitful" but of "no *direct* promise."[3] The female seedling discovered in 1977 had only 1 percent alpha acid content, but a male plant was later found in an unrelated breeding line, and within two breeding cycles a blueprint for a commercial dwarf variety was defined.

"This is why you keep the breadth of germplasm," Darby said. More recently he began reexamining records from a different angle. In addition to saving the germplasm he kept all the old literature, allowing him to look for varieties rejected because of aromas previously considered too intense but now of interest to brewers and drinkers. "The ones described as 'too strong' rather than being 'Manitoban,' " he said. "There's a lot good already out there. It needs to be rediscovered, almost."

Hop merchant Charles Faram, located in the Midlands, recently started its own hop development program. This includes the resurrection of old varieties and trialing and promotion of new varieties.

Descriptions from the early 1950s of two varieties now being revisited indicate they interested growers because of attractive agronomic properties. However, in 1953 a booklet from the Association of Growers of the New Varieties of Hops warned, "From the point of view of their wilt tolerance and other cultural characteristics these two varieties have consistently maintained a reasonably high standard; but there is no getting away from it, the brewers as a whole don't like them." Paul Corbett, managing director at Faram, has spoken to growers who "remember them as having a stink that was not appreciated by brewers, who at that time were brewing traditional English bitters and milds that were more reliant on Fuggle and Golding for their flavors. We believe that this is why development of the stronger flavors in the English breeding programs were avoided."

Darby understands the demand for new aromas. "People want impact hops. Everybody is looking to New Zealand and the U.S.," he said. "The brewers ask, why do they have to buy American when they want to buy English (grown) hops?"

He knew better than to predict what would come from several crosses he made with Cascade, including one with a wild Japanese hop. For another, he used the Russian hop, Serebrjanka, that is part of Cascade's lineage, thus "backcrossing" a hop that itself was the result of backcrossing (in that case, Fuggle with Fuggle). Plant breeders use the technique to introduce a particularly desirable gene or set of genes into a variety. "I'm looking for unusual flavors," he said. "I know we will get floral and citrus. What I am looking for is not that. It is something unexpected. Something unpredictable."

Advances in science aside, the unpredictable remains unpredictable. "You make the crosses on the conceptual level, but once they go out in there, it is a big melting pot," Darby said. "Every one is an individual."

Cascade has been particularly attractive for breeders. Lutz also used it for multiple crosses in Germany. Toru Kishimoto's studies in Japan indicate that it is rich in a black currantlike aroma that characterizes American hops, although without as much 4MMP as varieties such as Simcoe. Surprisingly, when brewers evaluated the hop before it was released in 1972 they compared it to Hallertau Mittelfrüh.

Val Peacock probably describes Cascade best, calling it a mutt because of its English-Russian-American background. USDA breeder Stan Brooks selected the seed for the plant 56013 in 1956, and a dozen years later it seemed it might simply end up in the germplasm library. No breweries showed any interest when the first test plots were harvested in 1969. However, after verticillium wilt devastated Mittelfrüh, Coors became the first brewery to support the new variety, offering contracts at then-lucrative prices, and soon other breweries also committed to using it.[4]

In 1976 Cascade accounted for 13 percent of hop acreage in the United States, but its popularity was short lived. "The hope placed in them originally as a replacement for imported European aroma hops was not fulfilled," the

1977 *Barth Hop Report* noted. Production quickly exceeded demand, and soon thereafter growers began taking Cascade out of the ground.

Steve Dresler was in his third year at Sierra Nevada Brewing in 1985 when he traveled to the Yakima Valley with brewery co-founder Ken Grossman for hop selection. "We go up there to buy, and we hear one of the major breweries, I don't remember which one, had pulled out of Cascade. Everybody told us it could go away. That's when we began contracting three years forward," Dresler said. Cascade production shrank 80 percent between 1981 and 1988 before it stabilized and began to rise again. Today it is the best selling American-grown hop used for aroma purposes. Sierra Nevada buys more Cascade than any brewery in the world, and the hop no longer depends on support from a few large companies.

"This whole hop world has gotten turned upside down. The holy grail in hop varietal development used to be to try and get a very large brewery to pick up and use your variety," said Eric Desmarais, a Yakima Valley farmer who has started his own small breeding program. He planted high alpha hops on two-thirds of 600 acres in 2009. Three years later he devoted 550 acres to specialty hops, most notably the proprietary variety El Dorado, to which he owns the rights.

"There are twenty new hop varieties growing in one of my hop lanes," Desmarais said, laughing loudly. "A lot of brewers like me are collecting (open pollinated plants). Seeing what they give us." It makes sense to Peacock. "That's hop breeding. That's how (landrace varieties) were selected to begin with," he said.

Desmarais didn't wait for whatever might show up on the side of the road. He hired somebody to make the crosses that resulted in El Dorado. "We'll let (the breeding details) out there at some point," he said. "It's completely different." He later made contact with Todd Bates, a New Mexico farmer who collected hops growing wild in Northern New Mexico. USDA research plant geneticist John Henning tested those hops and matched some with *neomexicanus* (wild American) hops in his collection. Others resembled natural crosses between wild hops and cultivated ones settlers would have brought with them. Bates since has crossed *neomexicanus* plants with each other, and in 2011 Desmarais planted two varieties from New Mexico on multiple hills. He understands the importance of agronomics and said that those hops needed at least another season to prove themselves in the field.

What Are Wild Hops?

• Original wild hops, which have never been grown commercially or used in breeding.

• Escaped original domestic hops, which might previously have been culti-vated in the vicinity of towns and monasteries where beer was brewed.

• New genotypes of wild hops, which arise due to mutual pollination.

He looks at hops through a farmer's eyes. What he saw in 2011 was a hop much different than anything else cultivated in Washington. The cone-to-leaf ratio is higher, the nettles on the bines are much larger, the laterals grow differently, and the leaves are a very dark green and almost waxy. "During the growing season, I have consultants walk the fields on a weekly basis, doing disease and pest scouting," Desmarais said. "These guys have been walking hop yards for 18 years, and walk most of the U.S. hop industry yards. They see just about everything. I didn't tell them what these were last year on purpose, to see what their reaction was. They knew they were looking at something very different."

Brewers who visited the Desmarais farm during harvest also noticed how unique those hops are. "They have a different aroma than anything else I've encountered. People (who smelled the 2011 crop) are intrigued," he said. "There are 2,000 craft brewers out there and 2,000 different opinions. If it fits agronomically, let the baby out. That's the direction I am going, rather than have me and a select few decide what brewers want. You have to let the customer decide."

He learned that with El Dorado, a hop he initially thought would be much too bold for brewers. In 2012 he licensed it to a second grower in another part of the Yakima Valley, expanding production as quickly as physically possible. He has a hard time finding adjectives to describe exactly what sets El Dorado apart, but he sees attributes similar to other popular varieties. "They are more intense. They all fit in this zone, 9 to 12 percent alpha acids, 24 to 32 percent cohumulone, high total oil. Those kind of hops that explode at you," he said. Brewers may not completely understand essential oils, but many are sure they want more.

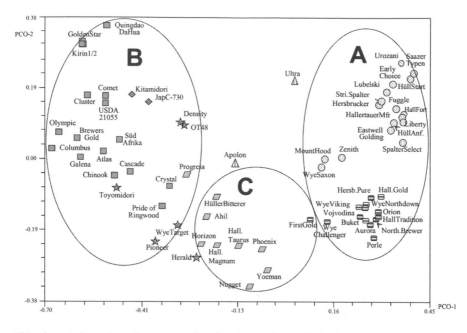

This chart is based on the results of molecular studies published in Plant Breeding *in 2000 and indicates the genetic distance between certain varieties. Group A represents mainly hops of European origin. Group B includes varieties of interest because of higher alpha acid content, their ancestors a mixture of European cultivated hops and North American wild hops. The hops in Group C resulted from crosses between hops of European and North American origin made with the objective to combine higher alpha acid content with European aroma notes. Courtesy of the Bavarian State Research Center for Agriculture.*

Interest in oils has changed in the years since Henning began working at the USDA in 1996. The spread of powdery mildew prompted the government to start the program in 1930, at a time when Oregon farmers grew 50 percent of the hops produced in the United States. The cultivation of Cluster moved to the drier states of Washington and Idaho, while Oregon growers stuck with the more resistant, but lower yielding, Fuggle. Research first focused on evaluating hops from England and Europe for domestic production, cross breeding began in earnest in the 1950s, and Brooks collected germplasm that eventually resulted in the release of multiple new varieties.

More than 80 percent of the hops Oregon farmers grew in 2011 were bred by the USDA. Washington growers planted substantially more from private breeding programs run by hop merchants, as well as varieties developed at Washington State University. Henning broadened the scope of the USDA program, incorporating more research related to hop genome sequencing, molecular markers, and genetic mapping. Using the markers, he has identified lines that may be more disease resistant.

Among the cultivars in USDA trials in 2012 was one that likely wouldn't have advanced beyond the first field a few years before. It smells of chocolate. Henning made the cross in 2003 as part of a large experiment where he chose parents that were either genetically similar or genetically diverse based upon molecular marker fingerprinting. He evaluated the plant on a single hill from 2005 until 2009. It advanced to multiple hill plots at multiple sites in 2010; on the USDA Agricultural Research Service farm, under the sponsorship of the Oregon Hop Commission at a commercial grower's yard, and on a plot at Washington State University.

"Several more years of evaluation will be pursued under these conditions unless a significant demand arises based upon actual pilot brewing," he explained after the 2011 harvest. "If it still passes muster at this point, it will be grown out in larger scale plots, say, 30 hill plots, under commercial conditions to see how it handles under normal agricultural practices. This would take another additional three years. Thus, it'll be at least six years down the road before release, unless HRC members decide they'd like to see it released sooner."

The largest breweries, who naturally bought the most hops, previously made most of the decisions. "In years past it was, 'What does Anheuser-Busch want?'" Henning said. "Other brewers had to speak separately. You'd talk to Anheuser-Busch, talk to Miller, talk to Coors. Each one of them had their idea of the hop they wanted. With A-B sometimes it would be a decade and they'd finally say, 'Yes, this will do.' With craft brewers it's, 'Great, we'll use it now.'"

"Some want more citrus, or more tropical, to mimic the flavor of Citra. Some just want more oil, as much as you can give them, as long as it is decent."

Conversations about oil are abundant. Consensus is not. "I am selecting for lower alpha, because craft brewers have told me they want low alpha," said Shaun Townsend, who heads the Aroma Hop Breeding Program started by Indie Hops in 2010. "I'm spending more time with high oil, because they also want that."

He's not certain if he can decouple alpha and oil content, but sometimes the sideways view, a wine tasting term for tilting the glass to get a unique look at the edge of the wine itself, reveals something altogether surprising. In Kent, Darby asks one of those questions for which an answer is not immediately evident. "It's only an opinion, but it seems oil chemistry and resin chemistry are much more interrelated than (we) currently believe," he said. "In going for high alpha you are changing the balance of oils. (Right now) I can't put facts and figures to that."

Gene Probasco, hop breeder and vice president in charge of farm and agronomic services at John I. Haas, started the first private breeding program in the country at Haas in 1978. For years when brewers talked about desirable aroma they used adjectives like "mild" and "noble." "Mild aroma was associated with low oil, and higher oil meant strong aroma," he said during the 2011 harvest, considering the question about the link between alpha and oil. "I've never seen high oil and low alpha. That would be hard, because to get oils you have to have high resin, and usually that means high alpha," he said. Still, he wouldn't rule it out. "Some oil glands are not part of lupulin, in theory."

Anton Lutz viewed the question from another angle. "So far, we cannot say. But we are not fixed on the amount of oil," he said. "We are interested in the composition."

Townsend has been dealing more closely with craft brewers since he started the Indie Hops breeding program. Also an assistant professor and senior research associate at Oregon State, he previously worked at the USDA with Henning. They continue to collaborate on identifying and adapting new molecular biology technologies to hop breeding. Townsend invited brewers to join him in the field when he was selecting hops at harvest time in 2010 and 2011.

"The thing that struck me is, there was hardly any agreement on *what* they smelled like," he said. "But they did agree on what they liked. The acid test is, 'How does it brew?' But if it smells good to you in the field it's worth pursuing."

Brewers told him what they are looking for right now. "They just want something different. Not lemon-citrus," he said. "They want tropical fruit, melon." Those who mention "something different" often quickly add "like Citra," citing a variety that epitomizes the concept of "flavor hops."

Triploids: No Seeds, New Flavors

Operating a breeding program that is basically independent of others in the world, R.H.J. Roborgh in New Zealand developed triploid varieties that laid the foundation for ones, such as Nelson Sauvin and Motueka, that fit in the "special" category. Almost all hops plants are diploids, with 20 chromosomes. Tetraploids, with 40 chromosomes, may occur naturally but most often result from treatment in a laboratory. When a breeder crosses a female tetraploid with a male diploid, the resulting triploid will be basically seed free, which brewers prefer. Roborgh's successor, Ron Beatson, has been particularly adept at creating hops with relatively high alpha that retain oil profiles much like their European predecessors. Most also include Cluster and perhaps some other American wild hops in their background, certainly a blend of Old and New World parentage that is reflected in their aroma and flavor.

Al Haunold at the USDA in Oregon released the first American triploids in 1976—Willamette and Columbia, both daughters of Fuggle. Willamette became the most popular "aroma" hop grown in the Northwest, primarily because of Anheuser-Busch, and farmers abandoned Columbia. Haunold made crosses for many other triploid cultivars—including Mt. Hood, Liberty, Crystal, Ultra, and Santiam—that he or John Henning, his successor, later released. They were designed to replicate traditional European aromas as closely as possible, unlike in New Zealand, because that's what brewers asked for at the time.

Probasco made the cross in 1990 that resulted in the Citra seedling. At the time brewers didn't talk about what would later be called "special" aroma, but "that's where all the interest seems to be these days," he said. In 1990 he cross-pollinated two plants, a sister and brother that resulted from a 1987 cross between a Hallertau Mittlefrüh mother and a male from an earlier cross.[5] "The cross was made for aroma," he said.

It was part of a project for a brewery client, one that lasted three years and created 150 potential cultivars. The brewery produced single-hop beers with each of them, and Probasco tasted them all. The one brewed with a hop called X-114 stood out. "I recognized that as something special," he said.

Nothing resulted from that project, but a few years later another large brewer contracted with Haas to develop a hop with unique aroma. Along with new crosses, Probasco gave them X-114. They brewed a test batch. "It tasted exactly like it did the first time," he said. Again, nothing resulted, but he kept the hop alive on a seven-hill plot he calls the museum. "I never forgot about that flavor."

Sometime after 2000 he was traveling with Pat Ting, then a hop chemist at Miller Brewing.

"I need a citrus variety," Ting said.

"I have one," Probasco replied.

The next year he shipped a two-pound sample to Miller, which since has merged operations in the United States to become MillerCoors. Miller provided the financial support for the first commercial production of Citra. "They tested it for quite a few years," Probasco said. Miller basically owned the hop for two years. Troy Rysewyk brewed a batch called Wild Ting IPA, dry hopping it with only Citra, in the pilot brewery. "It smelled like grapefruit, lychee, mango," Ting said. "But fermented, it tasted like Sauvignon Blanc."

In 2002 Haas and Select Botanicals Group merged their breeding programs to form the Hop Breeding Company. HBC began sending breweries samples, including small cuts of X-114. Eventually Widmer Brothers Brewing and Deschutes Brewery in Oregon, and Sierra Nevada Brewing in California, financed more acreage. In 2008 Widmer Brothers Brewing won a gold medal in the World Beer Cup with an American-style pale ale hopped with Citra and called *X-114*. Eighteen years after its parents were cross-pollinated, Citra was an overnight sensation. "We've been expanding it as fast as we can," Probasco said, talking about the limited number of farms that grow the hop. "We still don't have enough."

The process was educational in other ways. He tasted more than 100 single-hopped beers, each showcasing the character of a different cultivar. "It was a real eye opener. Some of the beers were just bad," he said. "Some of them were very, very good. Some varieties had a nice smell and they made bad-tasting beer. There weren't many that smelled bad, but some of those made good beer."

When Lutz explains the logic behind breeding a hop like Herkules— the combination of high alpha, good yield, disease resistance that results from an intentional combination of females and males—it's easy to forget that particularly popular varieties are literally one in a million, chosen from that many seedlings. Breeders also find aroma a far more elusive target than alpha. "We don't want to let anything unique get away," said Jason Perrault, who like Probasco carries many business cards. In addition to breeding hops he's in charge of sales for Perrault Farms.

The Patented Aroma of Amarillo

Darren Gamache was not having a very good September day. He was headed to show his father, Virgil, a letter he had drafted to inform Gamache Farms customers that the 2011 harvest did not produce the amount of Amarillo hops they had forecasted, and many brewers would not get as much as they ordered.

Amarillo is unique among proprietary (patented) hops, because it was found in 1998 in a field in the Toppenish region of the Yakima Valley, as opposed to emerging from a breeding program. "I'm surprised it exists. Every year we send crew out to rogue out the off types," Gamache said. The hop Gamache Farms eventually named Amarillo not only survived but stood out. It was not at all like the Liberty, noteworthy for its mild, traditional character that it was growing beside. Amarillo has bright, citrus aromas that are more intense than Cascade or Centennial. Its lineage remains unknown. Gamache, who was in high school at the time, picked it by hand. "My father smelled it and said that's what a beer should taste like," Gamache said.

"We gave it away for years," he said. However, demand eventually outstripped the number of acres the family could harvest in a timely manner.

Gamache is a fourth-generation farmer on his father's side, fifth-generation on his mother's. He remembers planting Tettnanger hops, part of an Anheuser-Busch experimental program, when he was seven years old. There are glasses inscribed, "Anheuser-Busch Hop Team 1988-1992," in his office.

"We're interested in new flavors ... but breeding programs are frightfully expensive," he said. Instead, beyond Amarillo the Gamaches grow varieties many others in Washington don't, such as Challenger, Golding, and, most notably, Sorachi Ace. In 2002 Gamache came across the hop at the USDA germplasm repository at Prosser, which contains varieties available to all farmers, and requested roots to grow the next year.

"We'll find a way to make it work," he said, returning to the topic of Amarillo and acknowledging one farm could no longer meet demand. Months later he struck deals to license other farmers to grow Amarillo, and by February he was making cuttings. At that moment he had a different view of hops and romance.

"Ask my wife about digging roots on Valentine's Day," he said.

In January 2012 HBC shipped small samples of more than a dozen experimental varieties to 10 breweries around the country. Employees at each brewery rubbed and sniffed the hops and rated them from one to seven relative to numerous descriptors, such as herbal, citrus, tropical fruit, and onion/garlic. The breweries returned the results to HBC. "The marketing and breeding programs used to be a lot more behind the scenes," said Perrault, and breeders worked with only a few large breweries. "For them it was like turning a giant ship around in the ocean," he said. "You can send (craft brewers) a sample, and they'll have a beer in a few weeks."

For instance, HBC 369 established a reputation before Probasco and Perrault named it Mosaic in 2012 and propagated enough plants to begin commercial production. Mosaic is a daughter of Simcoe, a hop that Select Botanicals Group (SBG) owns the right to.

Perrault was in high school in 1988 when he helped Chuck Zimmerman cross-pollinate several males and females of interest. "His feeling was it biased things talking about pedigree," Perrault said.[6] Zimmerman, who previously ran the USDA research facility at Prosser Station, kept a nursery at his house. "Warrior and Simcoe came out of those crosses. Simcoe was in row 2, hill 56. I still remember that. It had a nice-looking growth in the field."

Simcoe is susceptible to powdery mildew and yields fewer pounds per acre than many varieties. Its future was still uncertain in 2001, when Perrault crossed Simcoe with a Nugget-derived male bred for resistance to powdery mildew. Some of the resulting seedlings had Simcoe's unique aroma attributes plus others of their own, and one numbered 369 also turned out to be disease resistant with a much better yield. After making test batches with it, several brewers commented on a distinctive blueberry note in the aroma.

Production of Simcoe increased almost four times between 2010 and 2012, and the farmers growing it couldn't keep up with demand. Its future didn't look nearly as bright just a few years before. Following an initial rush of interest a few years after Yakima Chief Ranches (the breeding company that became SBG) released the variety in 2000, Perrault considered taking it out of the ground. "The demand just wasn't there," he said, remembering the amount of unsold inventory built up by 2006.

A few days before he recounted the history—how the attention beers such as Russian River *Pliny the Elder* and Weyerbacher *Double Simcoe IPA* sustained interest until a few larger breweries put in substantial orders—he was seated at a bar in Yakima. "Somebody came in and asked the bartender if he had any beers with Simcoe in them," Perrault said. "As a breeder and a grower it's fun to talk to brewers and drinkers about varieties."

He certainly has their attention. In September 2011 Lagunitas founder Tony Magee praised him on Twitter, writing, "Who would u guess 2 be the most important person in US craft brewing? A brewer..? Think again. He's a Hop grower named Jason Perrault! Word."[7]

Perrault was a bit embarrassed. "I think what his comment really represents is this relatively new appreciation for the potential impact of hops in beer flavor and aroma. Brewer innovation often comes in the form of creating new beer styles or enhancing existing ones; hops are now part and parcel of that," he said. The front end has changed. The back end has not. He and Lutz speak with almost the same meter when explaining the process that starts with 100,000 seedlings, the decisions that get made the first spring, the first fall, the next spring, across 10 years or more before perhaps a single plant ends up in a commercial field.

"It can be a depressing job," Perrault said. "You are always looking for the negative."

Lutz has a more colorful description. "A breeder is a mass murderer," he said.

A discriminating one, because the rules have changed. Townsend has seen that since 1999, when he began his post-doctoral work in Corvallis. "John (Henning) and I, we had some beautiful, beautiful plants," he said, discussing varieties that were rejected. "Something was not quite right for the larger brewers."

And You Thought Rabbits Only Liked Carrots

Occasionally a hop grower will ask the USDA in Corvallis to use DNA finger-printing to assure a hop variety is true to what it should be. While walking through test fields where they work, Peter Darby and Anton Lutz both told stories that suggest a simpler test might sometimes work as well.

Standing in a Kent field and talking about how Cascade differed from Fuggle, the English hop that is an important part of its parentage, Darby said, "Here, Cascade is very attractive to rabbits. We see that they go for the seedlings."

Pointing to blue plastic protection wrapped around the base of some plants at Hüll, Lutz explained that rabbits seemed drawn particularly to aroma types. They provided an early indication Herkules was not a typical high alpha hop, heading to the young plants the first chance they had. "When we planted it out for trials we said, 'We are feeding the rabbits,'" Lutz said, grinning widely.

Germans long ago included germplasm descended from American wild hops in breeding, but the use of Cascade as a mother only recently added a novel twist. Cascade was part of crosses that resulted in three of the four varieties the Society for Hop Research released in 2012. They smell more like German hops with an American accent than American hops with a German accent. "We have to find our own way. We come from a different place," Lutz said. "We want it balanced."

Lutz would define balance much as John Keeling, brewery director at Fuller's in London, would. "To have balance in the beer does not mean simply to go to the middle, bland flavors," Keeling said.

In the past Hüll released varieties intended for a broader farming and brewing audience, only choosing those with the potential to be planted on 1,000 or more hectares (about 2,500 acres). "For these (new ones) we must expect they might grow only 20, 50, 80 hectares," Lutz said. "And maybe a new variety is good for only five to eight years."

Per usual, he was thinking ahead. "More varieties," he continued. "More change. More intense flavors."

Notes

1. During the course of travels, I also heard Martin Ramos, the ranch manager at Segal Ranch, and Oregon farmer Gayle Goschie referred to as "hop whisperers." There surely are more.

2. One way that those involved in the commerce of hops could "hedge" their positions was to bet on the annual yield. The hop duty provided the information needed. In *The Brewing Industry in England*, Peter Mathias wrote, "Periodicals carried regular reports of these yields, and odds were quoted in every paper through hop growing and hop marketing regions, over which large amounts of money changed hands."

3. Peter Darby, "Hop Growing in England in the Twenty-First Century," *Journal of the Royal Agricultural Society of England* 165 (2004), 2. Available at *www.rase.org.uk/what-we-do/publications/journal/2004/08-67228849.pdf*.

4. A more complete story about how Cascade made it into production can be found in *IPA: Brewing Techniques, Recipes, and the Evolution of India Pale Ale* (Brewers Publications, 2012), pp. 147-150.

5. The male crossed with Mittelfrüh in 1987 was grown from a seedling selected from a 1984 cross between U.S. Tettnanger (likely a Fuggle) and USDA 63015, a male hop that is also a parent of Nugget. Citra contains 50 percent Mittelfrüh, 25 percent U.S. Tettnanger, 19 percent Brewer's Gold, 3 percent East Kent Golding, and 3 percent unknown (or "mutt").

6. Darby has records at Wye that show Salmon listed hops on his breeding charts under aliases, because he knew farmers and brewers would be immediately biased if they knew their true heritage.

7. *https://twitter.com/#!/lagunitasT/status/11834687289622937. Retrieved 23 August 2012.*

4

Growing Hops

You don't meet many first-generation hop farmers

They are the children and the grandchildren and the great-grandchildren of hop farmers, and farmers themselves. Not everybody who grows hops these days was born into the business, but sometimes it seems like it. "The intensity of harvest brings out this *spirit*," said Gayle Goschie, whose grandfather began growing hops in 1905. She is president of Goschie Farms, which occupies about 1,000 acres in Oregon's Willamette Valley, 350 of them devoted to hops. "Night and day. It can be frantic. It's my favorite time of year." She remembers playing hide-and-seek in piles of hops and around bales, and that she didn't like every single thing. "The machine part was noisy, and the hops scratched," she said. "But there were always jobs to be done, even for a 10-year-old. Grab a broom."

"The biggest focus on any hop ranch is harvest, and when you are 8 that is pretty cool," said Eric Desmarais, a fourth-generation farmer born in 1968 in Washington's Yakima Valley. "The mid-'80s was a difficult time for hop farmers. My mom and dad did everything they could to discourage me, but since I was 13 I knew this is what I wanted to do."

Before Florian Seitz took over the family farm outside of Wolnzach in 2008 he studied agriculture at the University of Applied Sciences Weihenstephan, did his practical training at the Hop Research Center at Hüll, and visited the United States, where, among other things, he helped John Henning of the U.S. Department of Agriculture cross-pollinate hop plants. For his diploma work he examined 14 varieties of poplar trees suitable for turning into wood chips.

Like Goschie and Desmarais, he is a farmer first. His younger brother, George, became a brewer. Florian Seitz manages about 170 acres and grows about 50 acres of hops. His great-grandfather first planted hops on the land he now works in 1869. His father still takes charge of drying hops right after they are picked, a skill most prized by growers and brewers and constantly in short supply no matter the country. "He has a great feel," Seitz said—illustrating, of course, by squeezing an imaginary cone between his thumb and fingers. The Germans have a saying, "The hops want to see their master every day." That was more practical when farms were smaller, but Seitz's uncle, who is retired, walks the fields almost every day from May to harvest.

"We don't get many visitors," Seitz said, happily guiding his car across deep ruts in order to show off a drip irrigation system not common in the Hallertau. German growers maintain that because their farms are smaller they can more easily adapt to brewers' changing demands. But unlike in the Yakima Valley, farmers do not expect a full crop in the first two years after they plant a new variety. An irrigation system not only reduces year-to-year variations but could shorten the time before new plants are fully productive. Seitz planted Hersbrucker hops for the first time in 2011, and, presumably because of the irrigation, they yielded almost 70 percent of what he would expect from mature plants. (He later emailed photos of his father smiling and holding the "babies" after harvest.)

Although German brewers may trace the hops they buy directly to the yard where they were grown, knowledge does not necessarily flow the other way. Seitz only occasionally knows which beers his hops end up in. "As a grower you are proud when you see what happens with your hops, when the product made from your own product is good," he said. "For the brewers it is good to see where your products are grown."

After the 2011 harvest American brewer Florian Kuplent visited the farm. Seitz and Kuplent met earlier in the year in the United States, and Kuplent contracted to buy half of the Mittelfrüh that Seitz had planted. They since have talked about other varieties, including those recently developed at Hüll.

Growing Hops at Home in 10 Easy Steps

1. Check the map. Hops grow best between latitudes 30° and 52°. They must have 15 hours of daylight, 120 days without frost during the growing season, and six to eight weeks of dormant time with the temperature below 40° F (4.4° C). They can be grown outside this zone, particularly on a small scale, but yields generally will be lower.

2. Buy rhizomes. Both homebrew stores and nurseries stock hop rhizomes. Many vendors sell them through the mail and will be able to offer advice about what grows best, and most easily, in various climates.

3. Site the hop yard. A south-facing wall on a two-story building that gets six to eight hours of sunshine a day is perfect. Ideally, string a top wire about 18 feet high, then attach strings the hops will climb. As an alternative, erect a single pole, which is what most commercial yards did until well into the nineteenth century, and run strings to the top. (Of course, another choice is to install a complete trellis system.) The string needs to support plants that will weigh 20 pounds when mature.

4. Prepare the soil. Hops will grow in a variety of soils but prefer mildly acidic, with a pH between 6 and 7.5. It should be loose, so the roots may burrow deeply, and drain well.

5. Review the stages of growth. They are dormancy, spring regrowth, vegetative growth, reproductive growth, formation of cones, and preparation for dormancy. In future years, when the first shoots come out of the ground, you will want to trim them, then follow the same regimen as a commercial grower.

6. Plant the rhizomes. The sprouts should be facing up and the roots down. Build a mound about a foot high, which will aid drainage. They should be spaced three feet apart. The cones will grow on sidearms as the plant grows.

7. Train the bines. As they begin to grow, train two to three around the string. They will wrap clockwise and follow the sun. This is the time to add nutrients such as nitrogen and calcium.

8. Water, monitor, repeat. The plant roots want to be wet but not water-logged. Goschie Farms monitors moisture at one, two, and three feet deep, because the plant takes water from the first two feet. Moisture three feet deep is lost. Trim the bottom of the plant and pull any weeds. This will make it easier to inspect. As the plants grow it is important to monitor for diseases (below) and for insects.

9. Pick and dry the hops. Don't expect many, if any, cones the first year, or perhaps even the second. They will come off easily if they are ready to pick and should spring back when compressed. They should be picked before they shatter. It may be easier to cut the bine and lay it horizontally for picking. However, the plant is in the process of preparing for dormancy, sending the roots nutrients it can use the next year, so leaving the bines hanging will increase yield for the following year and also allow cones that are not quite ready to ripen. Otherwise, when cutting the bine leave some green matter standing. The hops should be dried quickly or they will rot. A food dehydrator or window screens, allowing for good air flow, should work well.

10. Prepare for next year. Don't cut down the last green matter until the first frost, then prune bines to a few inches and cover with mulch. In the spring, remove the mulch and open up the mounds and remove the first gamey growth.

Hop Diseases & Pests

Downy mildew

The disease was first spotted in Japan in 1905, in American wild hops not long after, and in Europe beginning in 1920. The damage caused by both downy mildew and powdery mildew was an important reason why American production shifted from the East to the drier West and Northwest. It is caused by the fungus *Pseudoperonospora humuli.* The disease first appears in the spring as infected shoots emerge. Infected shoots are stunted, brittle, and lighter in color than healthy shoots. Infected shoots are unable to climb. Flowers often become infected when blooming occurs during wet weather and young cones stop growing and turn brown. Infected roots and crowns may be completely rotted and destroyed.

Powdery mildew

Descriptions of mold from as early as the seventeenth century likely refer to powdery mildew. It is caused by the fungus *Podosphaera macularis* and remains a serious problem in the Pacific Northwest. It appears first as powdery white colonies on leaves, buds, stems, and cones. Affected cones become reddish-brown as tissues die or may turn medium brown after kiln drying. Under cloudy, humid conditions at 65° F (18° C) the fungus can complete its life cycle in as little as five days.

Verticillium wilt

It is caused by two related fungi, and the nonlethal strain is more common in the Pacific Northwest. Wilt caused severe problems in England and on the Continent in the 1950s, wiping out the breeding yard at Wye College and later decimating Hallertau Mittelfrüh fields. The lethal strains cause rapid death of leaves, side arms, and the plant itself. Symptoms of the nonlethal variety include yellow veining of the leaves and wilting of leaves and vines.

Hop stunt viroid

The subviral pathogen does just what its name implies: It stunts the growth of plants and can reduce alpha acid yield by as much at 60 to 80 percent per acre. It spread throughout Japan in the 1950s, and the extension center at Washington State University confirmed its presence in North American-grown hops in 2004. Symptoms of infection may not appear for three to five growing seasons, which increases the danger of the propagation and distribution of infected plants. It is viewed as an increasing threat.

Hop mosaic virus

The virus is one of three carlaviruses, but the other two do not cause any obvious symptoms on commercial hop varieties. On sensitive varieties, mottling may occur on the leaves, and severely affected plants may have weak bine growth and fail to attach to the string. Golding hops or those with Golding heritage are most susceptible. The virus reduces growth and therefore yield.

Hop aphid, *Phorodon humuli*

The hop aphid is small, 1/20 to 1/10 inch long, and may be winged or wingless. It causes the most damage by feeding on developing cones, which turn brown.

It secretes large amounts of sugary honeydew that causes a sooty mold fungi on leaves and cones, reducing productivity. It may also transmit plant viruses.

Spider mites, *Tetranychus urticae*
Smaller than aphids, just 1/50 inch long, spider mites also suck plant juices from cells. A minor infestation causes bronze leaves, while a severe one results in defoliation and white webs. Spider mites are most dangerous during warm, dry weather and not usually a problem for well watered plants.

California Prionus beetle, *Prionus californicus*
The large brown beetles are 1 to 2 inches long, but the cream-colored larvae, 1/8 to 3 inches long, cause the damage. They feed on plant roots, and a severe infestation can completely destroy a crown and kill the plant.

Kuplent grew up not far from Munich and began his brewing education in Bavaria. He worked in Belgium and England, then for Anheuser-Busch, before he and partner David Wolfe opened Urban Chestnut Brewing in 2011. He knows firsthand how rarely brewers visited farmers in years past (or, for that matter, do today). "Sometimes I hear brewers say they are too small to go talk to growers or for contracts," he said, shaking his head to show he disagrees.

Even before Urban Chestnut opened, he visited Goschie Farms. He recognized immediately the value of direct contact with farmers. "They want to know what we want. They are happy to grow the varieties that are not the most fun to grow when you have a long-term relationship," he said. "You both have the same goal."

Anheuser-Busch bought more than 80 percent of the four varieties the Goschies grew in 2008. But even before InBev took over A-B at the end of 2008 (creating A-B InBev) and began terminating contracts for Willamette hops in the American Northwest and for Hallertau Mittelfrüh in Germany, Goschie initiated conversations with smaller brewers, particularly those in Oregon.

"I was just intrigued by the industry, and here it was right in my backyard, an hour or two's drive away. They were talking wonderful things about hops. Businesses with that much passion about what they were doing, and in the ingredients," she said. "I thought, this is something our company can be part of."

She met Larry Sidor, then the brewmaster at Deschutes Brewery in Bend, in 2007 when the farm hosted a reception for those attending the First International Brewers Symposium. Goschie grew the first certified "salmon safe" hops (meaning they were raised following ecologically sustainable practices) in Oregon, and Sidor ended up including them in Deschutes *Green Lakes Organic Ale.*

"For the first time in our family's history, I had an opportunity to have a brewer sit down at the table and see them react to the hops I grew, to give me those real-life reactions," she said. "It's extremely important. Most of us want to have the interaction. To identify what has been successful. If there is an example of what we can tweak."

"I realized we can have those interactions between a brewer and grower. A true partnership."

That includes answering many questions. "These brewers really have a passion for hops, an appetite for so much information," Goschie said. "They want to learn everything they can twist and turn out of each handful of hops."

Each variety is different, but certain basics are universal. Hop plants are dioecious, meaning they may be either female or male. Farmers plant only females, from rhizomes, so each is genetically identical, because male cones contain little or no lupulin. Farmers in most parts of the world actively remove male plants, although pollination increases yield, because most brewers prefer seedless[1] hops and, perhaps as important, seeds present problems for processing pellets.

The plants are annual above ground and perennial below, wintering through an important period of dormancy before emerging with particular vigor in the spring. They may grow as much as a foot in a day. They are photoperiodic, and day length is a critical factor for both vegetative growth and flowering. While they grow between latitudes 30° and 52°, they thrive between 45° and 50°. For instance, Goschie Farms is located at 45°, the Desmarais farm at 46.6°, and the Seitz farm at 47.7°.[2] Days last about half an hour longer in the Hallertau than in Yakima Valley, and an hour longer than in the Willamette Valley.

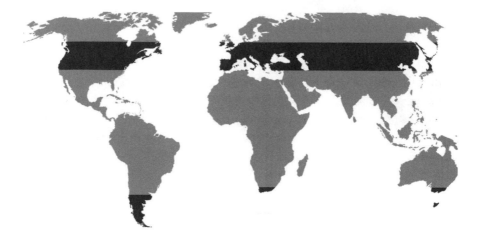

Hops grow best between the 35th and 50th parallels.

Not surprisingly, growing practices vary between regions and also within regions. Farmers in the the United States grow hops on 18-foot-high trellises. In Europe most trellises are seven meters (about 23 feet), but in Tettnang farmers hang Tettanger hops on eight-meter trellises and plant them closer together (although newer varieties grow on seven-meter trellises). Franz Wöllhaf, who works for the hop research station in Tettnang, said there are two theories as to why the narrower system evolved. "Some say they grow better if the wire is more straight," he said. "Others that it is because farmers had narrow tractors (because they also grow apples)." He shrugged and explained that Tettnanger hops grow better on eight-meter trellises because the space between laterals, the sidearms that extend from the main vine and produce cones, is larger.

Farmers plant 3,600 to 4,000 plants on one hectare (about 2.5 acres), training a single Tettnanger bine per wire, compared to two bines of other varieties. In Spalt Hans Zeiner trains two or three bines of Spalt Spalter to each of 4,000 wires in a hectare. At the Karel Dittrich farm in the Czech Republic workers train 3,300 Saaz plants in a hectare, putting two or three bines on a wire. In Yakima plants are usually spaced at 3½ feet by 14 feet or 7 feet by 7 feet, resulting in 889 plants per acre (the equivalent of about 2,200 plants per hectare).

Hop plants thrive in a wide range of soils, although there are varietal differences. The key requirements are sufficient depth of soil for the deep-rooted plant, adequate moisture, and good drainage. "A hop plant likes its roots wet and its head dry," said Kevin Riel of the Double R Hop Ranch in the Yakima Valley. In parts of England where soils are heavier, however, farmers worry that when a spring is particularly wet, plants may develop "wet foot syndrome," which inhibits development of root structure.

A new growing season in Spalt begins in March. Zeiner, who is manager of the Spalt hop growers association and also farms about 10 acres of hops himself, will open his hills and cut the first gamey shoots. He and his family train the second round of shoots to wire in May. Because Spalt farms are small, on average less than 10 acres, a family can rip out the other shoots by hand. Farmers tending to more acres use chemicals or dehydrate unwanted growth.

Hanging the twine or wire the plants will grow on, then training the plants to them, requires extra hands in the field. The Dittrich farm hires about 25 outside workers, many from Slovakia, in the spring, compared to 50 to harvest 104 acres. Seitz needs only three extra workers for harvest, but six during training, members of two migrant families. "One Polish family has been coming since the 1980s, and now their daughter works with them," Seitz said. Nearby, the Feiner family grows hops on about twice the acreage, around 107, and employs 20 migrant workers for training, compared to six at harvest.

"Every hop reacts differently, to the weather, to how they respond in the spring, when you put them on the string," Goschie said. Growing 12 varieties in 2011 compared to three just a few years ago sent her back to school. "It's a really good education. It forces me to pay that much more attention to detail, to rethink the traditional procedures we've done."

She learned Centennial is a difficult variety to train. "It just lays there," she said. Some hops climb better on twisted paper than coconut husk string. Other varieties almost train themselves, twining clockwise, following the sun. Charles Darwin literally watched hops grow while confined to a hospital bed, recording his findings in *The Movements and Habits of Climbing Plants*. "When the shoot of a hop (*Humulus lupulus*) rises from the ground, the two or three first-formed joints or internodes are straight and remain stationary; but the next-formed, whilst very young, may be seen to bend to one side and to travel slowly round towards all points of the compass, moving, like the hands of a watch, with the sun. The movement very soon acquires its full ordinary velocity," he wrote.[3]

Organic and Low-Trellis Hops

Organic hops and low trellises fit together well agronomically but also have a future independently.

The complexion of organic hop growing in the United States changed in 2010, when the U.S. Department of Agriculture announced it would remove hops from the Organic Exemption List effective at the beginning of 2013, eliminating an exception that allowed American brewers to use nonorganic hops in certified organic beer. In the two years after the announcement, organic acreage in the United States almost doubled. Equally important to brewers, the number of varieties available increased dramatically.

The future of organic hops depends ultimately on demand for organic beer, but the impact is already evident. American Organic Hop Grower Association president Pat Leavy pointed out, for instance, that organic production pushes the envelope, especially in breeding varieties that are less susceptible to pests, ultimately benefiting the entire brewing industry.

Except in New Zealand, which is basically pest and disease free, growing hops organically presents particular challenges that result in lower yields and higher prices. Because hop fields are so genetically uniform, they are more susceptible to pest and diseases, which conventional growers combat with frequent use of pesticides. The speed at which bines grow also makes supplying necessary nutrients, particularly nitrogen, harder for the organic grower. Plants require more water during intense growth, so controlling weeds that can affect that uptake is more important for hop growers and even more difficult than with other organic crops.

Low-trellis systems, known as hedgerows in England, reduce the environmental impact of conventional farming and also facilitate organic growing. Basically, low-trellis is the growing system (which may be used for nondwarf varieties as well as dwarfs), dwarf is the plant type (includes semi-dwarfs) and describes plants with a shorter internode distance than normal, hedgerow describes the dwarf type growing on the low-trellis system.

Obviously, plants only eight to 10 feet high (see color plate 1) are easier to monitor than those 18 feet and taller. Peter Darby at Wye Hops pointed out that because there is more plant material near the soil, and because hedges are in constant contact, dwarf hops attract more beneficial predatory insects, and it is easier for the insects to patrol the plants. Additionally, because bines are not cut and hauled to a separate facility for picking, they remain in place to provide nutrients as the plant prepares for the next season.

Hedge hops account for 25 percent of acreage in England, and Darby estimates that growers save about 45 percent on recurring costs, including labor. As important, English growers have access to true dwarf varieties. True dwarf hops produce adequate yields, while conventional hops generally yield between 25 and 80 percent less on low systems than high.

Dwarf growers may change varieties more quickly and economically, making it easier to meet ongoing demand for new and different aromas and for organic farmers to introduce greater genetic diversity.

The American Dwarf Hop Association is one of several organizations continuing research into low-trellis growing. Although Summit, planted extensively in the Yakima Valley, is not a true dwarf hop, it grows well on low-trellis systems and provides one of the few high alpha options for organic brewers.

Both public and private breeding programs worldwide now include dwarf hops. Using germplasm from England, John Henning at the USDA in Oregon already has several experimental varieties in the early stages of development. Like other breeders, Henning knows there is a bottom line.

"You still have to produce a hop the brewers want to use," he said.

The speed with which the plant makes its way up the string is forever captivating. "When they talked about Jack and the Beanstalk I'm convinced they were talking about hops," John Henning of the U.S. Department of Agriculture said.

Leaves grow from the nodes of the bine, which climb with the aid of hooked hairs called trichomes, as opposed to vines, which have tendrils. About the time the hops reach halfway up the trellis, laterals start to grow from the axils of the main leaf. Hop cones grow almost exclusively on the laterals, so their development ultimately determines yield.

Once farmers put the plant on string, vegetative growth takes off. Hops trained too early may reach the top of the trellis too fast, to the detriment of yield. Those trained too late won't grow enough. Researchers realized early in the twentieth century that day length controlled flowering plants, first describing the phenomenon known as "photoperiodic" in *Humulus scandens* and *Cannabis sativa*, establishing they are short-day plants, flowering as days grow shorter (after June 21 in the Northern Hemisphere).

The International Hop Growers Convention conducted a study in 1983 that illustrated the importance of growing cultivars with day length requirements suited to where they are planted. In the trial cultivars with a common female parent that had been bred and selected in England, Germany, or Yugoslavia, at latitudes of 51°, 48°, and 46°, respectively, were all grown in those three countries, and also in France at 47°. The cultivars flowered earliest in the lower latitudes, the difference between Yugoslavia and England being 10 to 14 days. The English and Yugoslav plants both showed a steady reduction in yield as the sites became more remote from their place of origin.[4]

The 2011 season in the American Northwest was a reminder of why English hop grower and broker Chris Daws likes to say that after the hops are put on string, the rest is "in the lap of the gods." Temperatures in the Yakima Valley were well below 50-year averages from May into August. Then it suddenly got hot. "It was perfect for cone development," said Riel. "It was the funniest maturing year I've seen in my life. The whole ranch was ready within a week's time."

Early maturing varieties were ready unusually late, but as farmers began to harvest them the heat arrived, and late maturing varieties were "ripe" early. Alpha acid levels rose quickly, but plant yields were down. For instance, alpha acids in Willamette hops grown in the Yakima region ranged between 7 and 9 percent, compared to a normal average of 4 to 5 percent. Centennial alpha acids reach 13 percent rather than 9 to 11 percent.

More than 100 years ago many English farmers grew several varieties of Golding hops that differed only in how quickly they matured. Farmers in the American Northwest grew both Early and Late Clusters. By staggering the maturity date of their plants in the field they extended the picking season, increasing the amount they could harvest without adding more equipment. Farmers do the same today. Some may make their decision about when to pick based on experience, others by using a formula based on dry matter remaining in a sample of cones.

"You want the cone to come off easily," Goschie said. If the cone is not mature, it will be difficult to pick, if it hangs too long it will shatter. Picking at the right time assures hops will dry in the optimal time and retain the whole cone. "They will come out nice and green, some more like golden," she said.

Increasingly, brewers are interested in hops harvested later. Zeiner understands: "The brewers want different hops, some want a greener hop, some want a hop that is a bit more mature. It is about the aroma, the aroma they want." He cautioned about waiting too long. "If it's too old the aroma is not so fine anymore. The aroma has changed, but I cannot say how it changed." In Spalt "*fein* (fine)" has a special meaning.[5]

Riel said Double R regularly picked the Willamette harvested for Anheuser-Busch four days before what would otherwise be the target date. "They wanted it treated differently and harvested differently," he said. "They didn't like a fully expressed aroma." For craft brewers we pick it one day later (than the target date). There are a lot more oils."

Research in Oregon, Germany, and Australia confirmed oils continue to increase at a faster rate than alpha acids after what is generally considered the optimal commercial harvest date. In addition, the study at Oregon State University concluded in 2011 found the composition of essential oils changed depending on harvest date. A sensory panel also noticed significant differences in beers brewed with hops harvested on different dates. For instance, a beer brewed using Cascade hops harvested late rated higher in melon and floral notes.[6] Likewise, in the German study both aroma intensity and quality increased the later hops were picked, and consumers showed a distinct preference for hops picked later.[7]

In 2008 Hop Products Australia began an ongoing study that included harvest dates and drying conditions. HPA determined that adding visual assessment of lupulin gland filling in preharvest analysis enhanced the chemical maturity of hops, and that aroma varieties benefited more. The study also concluded that lowering kilning temperatures from 140° F (60° C) to 122° F (50° C) increased the quantities of essential oils for certain varieties.[8]

Location, Location, Location

The OSU study also revealed something the researchers were not looking for, differences they didn't expect between hops grown miles rather than continents apart. They collected data for Willamette and Cascade hops harvested on three dates at three locations in the Willamette Valley. When farmers picked either variety early at any of the farms, the concentration of essential oils varied little. Differences began to arise as hops spent more time on the bine. They were not surprised that one location might yield hops richer in oils, but were when different farms proved "better" for different varieties. For instance, the amount of oils in Cascade hops picked at Farm 1 was lower than the other two, but the level of oils in Willamette grown on Farm 1 was highest.[9] The opposite was true on Farm 3. The German study focused on harvest dates also found location significantly changed hop aroma and flavor in the beers, even when the hops were grown only miles apart.[10]

Farmers learned long ago that most varieties grow better in a particular region and sometimes not at all in others. The OSU and German research provides confirmation that the differences extend beyond yield or levels of alpha acids, supporting both anecdotal and scientific discoveries. Eric Toft, brewmaster at Private Landbrauerei Schönram in Bavaria, has judged in a German national hop competition for several years. Panelists rate hops on aroma, appearance, and other criteria, honoring the best of each variety. Toft sits with farmers and breeders who have been judging for 30 years. They may rub and sniff only a few hops and tell immediately which yard they come from, because they've been there. The variety doesn't matter, although most farmers grow several. "They recognize the character of the place," Toft said.

Winemakers and wine drinkers might call that *terroir. Grand Dictionnaire Universel du XIXe Siècle*, Pierre Larousse's nineteenth century French dictionary, defines *terroir* as "the earth considered from the point of view of agriculture." It describes *le goût de terroir* as "the flavor or odor of certain locales that are given to its products, particularly with wine."[11] Even though local environments clearly influence character, the fact that wine writer Jamie Goode has described the the concept of *terroir* in wine as "blindingly obvious and hotly controversial"[12] is good enough reason to steer away from the term *hop terroir.*

The rolling hills around the city of Tettnang in southwest Germany are spectacular, covered with vineyards, orchards, and fields of hops, the Alps looming in the distance. Despite a strong genetic link to Spalt Spalter and Saaz, Tettnanger hops are different, both because of farming practices and *where* they are grown. When brewers such as Fritz Tauscher (p. 176) make beers that feature the hop, it seems a perfect fit for the terrain. However, Tettnanger hops also remain distinctive in a beer brewed thousands of miles away.

Henning explained the science behind why environment and epigenetics combine to make hops from a particular area unique. All plant species have methylated DNA, which causes some genes to be "switched on" more easily than others. Differences in soil, day length, temperatures, amount of rainfall, and terrain all may influence the methylation process. The underlying DNA does not change, but the methylation pattern can be different.

Researchers at the University of Hohenheim in Stuttgart used AFLP fingerprinting to analyze the similarity of Tettnanger hop plants to other varieties in 2002, reaffirming other surveys that concluded that Tettnanger, Spalt Spalter, and Saaz hops are so closely related they may be grouped together as "Saazer hops."[13] The Tettnang region (47.7° latitude) receives both more rain and more sunshine than areas around Spalt (49.2° latitude) or Žatec (50.3°). Žatec is about 700 feet above sea level, Spalt 1,200 feet, and Tettnang itself is 1,500 feet (although fields range from 1,300 to 2,300). Not only are many of the trellises in Tettnang taller, but many plants are 80 to 100 years old, compared to 25-year-old plants in other regions.

Their alpha/beta and oil profiles are similar but not identical. *The Hop Aroma Compendium's* descriptions of aromas found in each of them reflect those differences:

Saaz: "Spicy, woody, such as tarragon, lavender, cedarwood, and smoked bacon."

Spalter: "Woody aromas … reminiscent of tonka beans and barrique, with slightly sweet notes of ripe bananas."

Tettnanger: "Woody aromas and cream-caramel components, such as gingerbread and almonds, predominate, combined with fruity blueberry notes."[14]

All the Osvald Saaz clones studied at Stuttgart could be more clearly distinguished from each other than the original Saaz could from Tettnanger and Spalter. Three of the clones were quite similar to the landrace Saaz, but Osvald clone 126 was much closer to Fuggle (or U.S. Tettnanger). Nonetheless, all Osvald clones grown in the region around Žatec exhibited very similar morphological traits and aroma components.[15]

The Bogensbergers are the last farmers in the Hallertau region to grow Nugget, a variety bred in Oregon for its (then high) alpha. "It has a much better aroma than 20 years ago," Florian Bogensberger said. "All the imported varieties, they are very strong when they arrive, but they change." Plants bred for Germany may also change. "You look at Magnum (introduced in 1993), you see the leaf is much different between the first one and the one we have now," he said.

He grew the distinctive American variety Columbus, which is pungent and sometimes catty, for several years. "When it came over it was rough, but it turned into something smooth," he said.

The differences aren't necessarily pronounced. Tony Redsell's Golding and Fuggle hops regularly win awards for their quality. "On that day three judges think we have the best aroma," he said, modestly. Quite honestly, he said, he wasn't convinced he could tell his own apart from those grown on another Kent farm, but he would have no problem picking out Golding hops grown elsewhere. In 2011 he began the application process with the European Union seeking Protected Geographical Indication for the hop.

"There is a different style of aroma in Kent." He held his hands to his nose, rubbed his thumb against his fingers in the manner that once again seems to be part of every hop farmer's DNA. "There's a citrus. It really comes on strongly."

Size Matters, But So Does Family

Hop farms in the American Northwest, particularly the Yakima Valley, and in Germany barely resemble each other in size, but in neither area will a visitor meet many first-generation hop growers. They are family run operations. The *average* farm in the Northwest (Washington, Oregon, and Idaho) harvests about the same quantity of hops as all of Belgium and is 12 times larger the average German farm. Seventy-five farmers managed a little more than 920 acres of hops in the Spalt region of Germany in 2010. That's two more farms than in all of the Northwest, where farmers planted 32,000 acres. The small-scale model works elsewhere. In Spain 240 farmers grow almost exclusively Nugget, tending to an average of six acres.

Leading Hop Producing Nations (2011)

	Aroma		Alpha		Total	
	Acres	Pounds	Acres	Pounds	Acres	Pounds
Germany	23,569	40,784,100	20,248	41,887,400	43,818	82,672,500
United States	10,361	12,125,300	19,655	52,866,308	30,016	64,991,608
China	1,433	3,527,360	12,889	26,455,200	14,332	29,282,560
Czech Republic	10,791	13,503,175	161	275,575	10,952	13,778,750
Slovenia	3,168	4,629,660	178	220,460	3,346	4,850,120
Poland	939	1,102,300	2,718	3,306,900	3,657	4,409,200
England	1,977	2,425,060	593	925,932	2,570	3,350,992
Australia	119	171,959	1,006	2,129,644	1,124	2,301,602
Spain	0	0	1,260	2,081,142	1,260	2,081,142
South Africa	0	0	1,216	2,012,800	1,216	2,012,800
France	892	1,219,144	156	200,619	1,048	1,419,762
Ukraine	1,184	1,139,778	381	273,370	1,564	1,413,149
New Zealand	655	859,794	284	407,851	939	1,267,645

Source: International Hop Growers Convention

In 2011 (removing all numbers related to China because it is relatively insular) Germany and the United States produced 85 percent of the world's alpha/bitter hops and 67 percent of the aroma. The Czech Republic sold another 16 percent of the aroma.[16] Scores of others countries grow hops, often just for their home market, but the United

States and Germany set expectations for price, quality, and variety. The challenge that presents in a country like Poland, where farmers don't have high alpha or unique aroma hops and brewers can easily find a wider variety nearby, are obvious. Nearly one-third of Polish hop farmers quit growing hops in 2010 and 2011, although almost 700 remain. They manage an average of less than six acres each.

The largest grower in Washington, Roy Farms, produces more hops annually than all but six countries. Roy cultivates more than 30 varieties on more than 3,000 acres, on high trellises and low trellises, organic and mostly otherwise. The company has four processing facilities in Moxee and Toppenish, using four Dauenhauer picking machines, two of them double-size, plus a field picking unit for its low-trellis hops. It has two pellet mills that turn dried cones directly into Harvest Fresh Pellets. That's one more pelletizing facility than in all of New Zealand.

The benefits of size are obvious in Yakima. The advantages of long history are obvious in Germany and the Czech Republic. A new first generation of hop growers in the United States hopes to tap into growing consumer interest in local products. As with organic hops, demand will determine the newcomers' success.

Some projects exist on a very small scale. One example: in the spring of 2012 a group of Bronx community gardens, including several in the New York Botanical Garden Bronx Green-up program and a Catholic parish, planted 125 Cascade plants. The Bronx Brewery planned to use the hops they produced in an "Urban Hop" beer.

Steve Miller of the Cornell Cooperative Extension provided expertise for the project in the Bronx. The New York State Department of Agriculture and Markets provided the funding to hire Miller in 2011, and since then Governor Andrew Cuomo has proposed legislation to create a "Farm Brewery" license that would provide a variety of benefits to breweries that used New York-grown barley and hops. Both the University of Vermont and the Cornell Extension Service offer support for the Northeast Hop Alliance, which was organized in 2001 and became a not-for-profit in 2010.

Its members understand the need to merge history and modern tastes. "This is not new to Vermont. We're trying to relearn the crop," said Heather Darby, an associate professor of agronomy with UVM Extension, which planted a test patch called the Vermont Hops Project as part of a five-year project to assess which varieties would be suited to the local environment.

Early American settlers found hops growing in the wild. After two Dutch colonists built the first commercial brewery on the southern tip of what is now Manhattan, a fellow countryman wrote, "Our Netherlanders raise good wheat rye, barley, oats, and peas and can brew as good beer here as in our fatherland, for good hops grow in the woods." Later in the seventeenth century, a resident of New Jersey reported, "The lands from the capes along the Delaware River would bear great crops of wheat and barley, and it would be very fit for hop gardens. ... Hops in some places grow naturally, but were hop gardens planted in the low, rich land, quantities might be raised to good advantage."[17]

Brewers needed more hops than could be collected in the woods, and the first hop gardens were planted in 1629. During the next 200 years farmers in almost every colony, and then state, attempted to cultivate hops. According to the 1850 U.S. Census, hops were grown in 33 of the then 35 states, although often in quantities so small they could barely be called commercial. For instance, Kentucky produced "five or six bales" in 1873.

By 1879 New York grew 80 percent of American hops. Four years later, *The Western Brewer* provided an industry overview: "It will be seen that in 1850 hops were raised in 33 States and Territories; in 1860, in 37; in 1870 in 36; in 1880 but 18 ... It remains to be seen whether California, Oregon, and Washington Territory will increase their production; or, in a few years, drop off as so many others have done. It is probable that New York will always remain the banner hop state."[18]

Instead, within 10 years the Pacific Coast produced more hops than New York, and by the time Prohibition began New York farmers grew less than 4 percent of the national crop. According to Daniel Flint, who grew hops in California throughout the second half of the nineteenth century, hop cultivation in California began in 1855 in Alameda County. Sacramento brewers, accustomed to using hops that had been shipped "around the Horn" of South America, were shocked by the difference. "When they first tried fresh, strong California hops, they used the same quantity, with the result that the beer was too bitter to use," Flint wrote. "Consequently, they began to reduce the quantity used for a brew and to mix them with the old, imported hops."[19] Flint later wrote that most of the stock on the Pacific Coast came from his hop yard, which he planted in 1855 with hops acquired in Vermont.

Cluster: America's Landrace Variety?

In 1971, the year before the USDA released Cascade to farmers, Cluster accounted for almost 80 percent of U.S. hop acreage (Fuggle, Bullion, and Brewer's Gold combined for about 16 percent). In 2011 farmers planted only 1.5 percent of their hop fields with Cluster, basically the same amount as Amarillo.

Cluster contains genes from both American and European stock and obviously resulted from the natural pollination of hop plants imported from Europe and American wild stock. Where and when that took place likely will never be determined for certain, but in a sense selection was much like the landrace varieties on the Continent and in England.

After farmers chose particular varieties they continued to make clonal selections, and Early Cluster and Late Cluster emerged. Early Cluster, more widely grown at the height of Cluster cultivation, likely resulted from a somatic mutation on Oregon or Pacific Coast Clusters around 1908. Because it matured 10 days to two weeks earlier, it became known as Early Cluster, and the parent eventually was called Late Cluster. Late Cluster is much like (although it grows more vigorously than) Milltown and similar Cluster variations grown in New York after Prohibition.

Decades later Ezra Meeker, a well known Washington farmer, proposed that a model Pacific yard would include one acre of Humphrey's Seedling, two acres of Cluster, and one acre of Canada Red. Humphrey's Seedling originated by chance in Wisconsin, and Canada Red, which had red bines, in Canada. Cluster likely was the variation that eventually became known as Late Cluster. Oregon's farmers didn't follow his suggestion, instead planting Fuggle as well as Cluster. Fuggle exports, which remained strong during Prohibition, helped Oregon become the nation's leading hop growing state. Washington sales did not surpass Oregon's until 1942.

Washington hops first flourished in the Puyallup Valley near Tacoma after J.V. Meeker planted them in 1866, curing them in a loft over his living room. His son, Ezra, planted four acres the next year and by 1891 grew 500 acres himself and had an interest in almost every other commercial hop operation in the Northwest. Years after he died, a tribute in the local newspaper described him as "hop king of the world." In 1892 hop lice devastated crops along the Pacific Coast. Meeker wrote:

"One evening in 1892, as I stepped out of my office and cast my eyes toward one group of hop houses, it struck me that the hop foliage of a field nearby was off color—did not look natural. ... (I) walked down to the yards, a quarter of a mile away, and there saw the first hop louse. The yard was literally alive with lice, and (they) were destroying—at least the quality. ... At that time I had advanced to my neighbors and others upon their hop crops more than a hundred thousand dollars, which was lost. These people simply could not pay, and I forgave the debt, taking no judgments against them, and I have never regretted the action. All my accumulations were swept away, and I quit the business—or rather, the business quit me."[20]

The next year, the *Yakima Herald* reported acreage in the Valley grew from 400 acres to a state-leading 2,500 acres between 1891 and 1893. Much was planted by French Canadians, who bought land from the Moxee Company, an experimental farm started by Alexander Bell and others. In anticipation of future settlement, the company irrigated 7,000 acres of land. Irrigation, of course, has been vital to the success of hop growing in the Yakima Valley, much of it on farms run by the offspring of those French Canadian settlers. In 2011 Washington farmers harvested 79.3 percent (by weight) of the hops grown in the United States, Oregon farmers 12.3 percent, and Idaho farmers 8.3 percent.

Northwest growers understand the new farmers are not conventional competitors. They will not be shipping hops to the world's emerging beer markets, nor likely will provide more than a small percentage of what local brewers need. "We are really encouraged by this," said Leslie Roy, president of Roy Farms and vice president of the International Hop Growers Convention. "Anything that raises the awareness of hops is a positive. The real danger is being relegated to a commodity by the large breweries."

Farmers in more than a dozen states outside the Northwest grew hops for commercial purposes in 2012. In some regions, such as Michigan, they formed alliances that provide not only education but also pooled resources for more efficient picking, drying, and hop processing. Growing has resumed in states such as Wisconsin and New York, where hops once grew widely. In other states they are new. Colorado State University planted its first experimental organic hops in 2004, and farmers across the state have since harvested both organic hops and non-organic hops. Additionally, AC Golden Brewing, a subsidiary of MillerCoors, gave away Chinook rhizomes for volunteers to grow at home. About 500 volunteers took rhizomes in 2011, and 125 harvested them. The rather modest 60 pounds went into a batch of *Colorado Native Lager*.

Until relatively recently several large British breweries kept their own hop farms. In 1997 Whitbread Brewing's farm became the Hop Farm Family Park, a multipurpose, 400-acre facility with rides, games, picnic areas, and many other attractions, including a hop museum and a large collection of oast houses. The Rogue Ales Micro Hopyard in Oregon's Willamette Valley, about 75 miles east of the Newport brewery, does not have miniature golf, but Rogue has used it to teach consumers that beer is an agricultural product.

"This whole thing is an accident," Brett Joyce, Rogue Ales president, said late one September 2011 afternoon, looking at a field full of pumpkins with recently harvested hop yards in the background. After the hop shortage of 2007 Rogue first considered growing hops. "We didn't really know much," Joyce said, candidly. Obviously, brewmaster John Maier knows how to use hops. He established a reputation long ago for lavish additions of hops in his beers. "And we knew they'd been readily available at four or five dollars a pound," Joyce said.

Rogue didn't start from scratch. "We opened up the Yellow Pages and found the Coleman family," he said. Now Rogue leases 42 acres from the Colemans. Wigan Richardson and Company first established the farm early in the twentieth century. The Colemans, who continue to farm other land, already had a trellis system in place, a picking machine, and kilning facilities.

The farm sits hard beside the Willamette River, and boaters can dock for a beer in the tasting room or on a deck that looks out on the hop fields. A few months after the 2011 harvest the Willamette flooded, stranding farm managers Natascha and Josh Cronin for nearly two weeks. Water covered the dormant hop plants for 11 days. Rogue entered the summer uncertain as to what impact that would have on the 2012 crop, which was expected to produce about 40 percent of the hops the brewery uses. "We may have to buy hops on the spot market, and these will be expensive," he said.

"People say we must have it made," Joyce had said months before, drinking beer made with hops grown a cone's throw away. "But we take all the risk. It's our money if the crop fails." Even under the best of conditions, he said, the Microyard hops are the most expensive Rogue uses. "But it sure is fun," he said.

Rogue's experience with hops inspired the brewery to also grow a portion of its own barley, on the Tygh Ranch, far to the north and east of the hop yard.[21]

"Hops give us a platform to try this," Joyce said. The farm is off the beaten path, but by the fall of 2011 was packed with visitors most weekends. The tasting room keeps regular hours and during the summer and harvest there are tours every day. Special events include fish bakes and musical concerts as well as educational presentations, such as one on beekeeping. The facility is available for weddings, and there's overnight accommodations at the Hop 'N Bed, a farmhouse that overlooks acres of hops as well as the processing facilities. Rogue invites local homebrewers to make batches of beer in its nanobrewery, located in a simple shed. The beers are served at Chatoe Rogue.

"We've been here three years," Joyce said, considering what has been added each year. "I think we'll look back in three more years and see a lot more. This is about the Rogue experience."

He's not afraid to use the "T" word (*terroir*) or make comparisons to wine. "We believe origin matters. I don't think consumers view beer as an agricultural product."

James Altweis of Gorst Valley Hops in Wisconsin knows that the farmers he works with will succeed only if brewers and their customers appreciate the value of Wisconsin-grown hops, but he's sure they expect even more. "Some people perceive local as having value. Local's great, but it can't be the only part of our plan," he said. "If a brewer doesn't see the improvement, then he's not going to pay the higher price."

Gorst Valley works directly with growers, supporting them throughout the process that starts with planning a yard and continues through building their own version of an oast house. For the 2012 harvest Gorst Valley began selling small-scale harvesters designed to mechanically pick hops grown on less than 10 acres. It takes a crew of six an hour to pick two bines by hand, while the new Bine 3060, operated by three people, processes 30 to 60 bines an hour. The Bine 3060 costs

$12,900, while a large-scale picker would run around $180,000 new and $30,000 used.

Once a grower harvests and dries the hops, he or she delivers them to the GVH processing center to be pelletized, packaged, and delivered. Gorst Valley acts as a broker and keeps a percentage of what a brewer pays. The price will be considerably more than the same variety of hops might cost from an established growing region. "The industry we are creating has no option to compete on price," Altweis said.

Wisconsin growers need look only to history to see the results when hops are traded as a commodity. When hop louse wiped out much of the New York crop in 1867, hop mania swept across Wisconsin, after hops that sold for 15 to 25 cents per pound in 1861 brought up to 70 cents a pound. Newspapers carried stories about the profits farmers made on every acre they planted. The accounts described the mansions the growers lived in, the carriages they rode in, and the vacations they took. Little surprise that production in Sauk County, where most of the state's hops grew, jumped from a half million pounds in 1865 to four million pounds in 1869. Hop prices crashed from 70 cents a pound in 1867 to four or five cents the next year. Both growers and hop dealers went bankrupt.[22] Wisconsin farmers were pretty much out of the hop business within the next 10 years, and the "Hop Crash of 1867" was their legacy.

Wisconsin craft brewers buy most of the hops Gorst Valley growers plant—the company processed about 100 pounds of hops in 2009 and expected 10,000 in 2012, still less than 20 percent of the production of an average farm in the Hallertau—and the rest go to craft brewers in other states or to homebrewers. About 70 percent of those are aroma hops. "That's what the brewers want," Altweis said. "We have to look for what we can do on process (drying and pelletizing) that adds value, that creates differences apparent in the final beer."

Much of that occurs after hops are picked. A farmer's work isn't done until he or she delivers dried bales to a processing plant.

Notes

1. Until 1976 English farmers planted males in order to produce hops that weighed more (sometimes up to a quarter of their weight was seed) and so sold for more, and except for varieties bred to remain seedless many English hops still contain seeds. In Europe the seed content of dried hops must be below 2 percent for them to be considered seedless. In the United States the limit is 3 percent (although that would be 4.2 percent using the European measurement method). Oregon is the only large hop region outside of England growing a substantial amount of seeded hop. About 36 percent of the 2011 crop contained 4 percent or more seeds (as compared to 9 percent of the Washington crop).

2. Farmers in New Zealand and Australia together do not plant two-thirds as many acres of hops as those in Oregon, but their hop aromas currently are very fashionable. New Zealand's Nelson hop growing region is at -41°, while two-thirds of Australia's hops are grown in Tasmania (-42°) and the others at about -36.5°.

3. Charles Darwin, *The Movements and Habits of Climbing Plants* (London: John Murray, 1906), 2-3.

4. R.A. Neve, *Hops* (London: Chapman and Hall, 1991), 18.

5. The local brewery, Stadtbrauerei Spalt, is owned by the town of Spalt. The mayor is also director of the brewery, and each of the town residents, about 5,000, may vote on major brewery decisions.

6. Thomas Shellhammer and Daniel Sharp, "Hops-related Research at Oregon State University," presentation at the Craft Brewers Conference, San Francisco, 2011.

7. B. Bailey, C. Schönberger, G. Drexler, A. Gahr, R. Newman, M. Pöschl, and E. Geiger, "The Influence of Hop Harvest Date on Hop Aroma in Dry-hopped Beers," *Master Brewers Association of the Americas Technical Quarterly* 46, no. 2 (2009), doi:10.1094/ TQ-46-2-0409-01, 3.

8. S. Whittock, A. Price, N. Davies, and A. Koutoulis, "Growing Beer Flavour—A Hop Grower's Perspective," presentation at Institute of Brewing and Distilling Asia Pacific Section Convention, Melbourne, Australia, 2012.

9. Shellhammer and Sharp.

10. Bailey, et al., 4.

11. Amy Trubeck, *The Taste of Place* (Berkeley: University of California Press, 2008), xv.

12. Goode, "Terroir Baggage," *Wineanorak.com*. Retrieved 23 August 2012 from *www.wineanorak.com/terroirbaggage.htm*.

12. R. Fleischer, C. Horemann, A. Schwekendiek, C. Kling, and G. Weber, "AFLP Fingerprint in Hop: Analysis of the Genetic Variability of the Tettnang Variety," *Genetic Resources and Crop Evolution* 51 (2004), 218.

13. Joh. Barth & Sohn, *The Hop Aroma Compendium,* Vol. 1 (Nuremberg: John. Barth & Sohn, 2012), 65, 73, 83.

14. Fleischer, et al., 217.

15. The IHGC classifies hops based on the alpha acids in varieties, so the percentages would be a little different if adjusted for the way some brewers use New World hops. That doesn't change the overall trend.

16. L. Gimble, R. Romanko, B. Schwartz, and H. Eisman, *Steiner's Guide to American Hops,* (Printed in United States: S.S. Steiner, 1973), 39.

17. Ibid., 52.

18. Daniel Flint, *Hop Culture in California,* Farmers' Bulletin No. 115 (Washington, D.C: U.S. Department of Agriculture, 1900), 9.

19. Ezra Meeker, *The Busy Life of Eighty-Five Years of Ezra Meeker* (Seattle: Ezra Meeker, 1916), 228-229.

20. Sierra Nevada grows both the hops and barley for its *Estate Homegrown Ale*. The brewery first planted three acres of hops in an adjacent field in 2002 and now manages eight acres, all grown organically. In full bloom the minifarm beside the brewery is quite striking, and larger than most of the new wave enterprises elsewhere, although it could appear more pastoral: Viewed from the north, a Costco warehouse looms in the background.

21. Gimble, et al., 46.

5

Harvesting Hops

*Where the violence of picking machines
meets the quiet of the kiln*

In the midst of preparing for his sixty-fourth hop harvest, farmer Tony Redsell took time late in the summer of 2011 to answer a phone call from a local newspaper reporter. He patiently and quickly explained the basics of the hop harvest, both past and present. "We're coming to the silly season," he explained after he hung up. "Every editor thinks, let's have an article on hops, 'Hopping down in Kent.' "

Redsell is the most famous of the 50 or so hop growers who are left in all of England, with his 200 acres of hops in Kent representing about 10 percent of the national total. Hop production in Kent reached its peak in 1878, when farmers grew hops on 77,000 acres. Special trains that ran between London and Kent carried 23,000 seasonal pickers to spend a moneymaking vacation picking hops, an annual pilgrimage that inspired the song, "Hopping Down in Kent."

The influx of workers was repeated everywhere hops were grown, until mechanical pickers made the harvest less labor intensive. In the west of England, the population of the village of Bishops Frome swelled from 700 to more than 5,000 throughout the 1920s and 1930s. In 1868 about 30,000 girls went to work in a single Wisconsin hop-growing area, 20,000 of them from other parts of Wisconsin. As recently as the 1950s, city officials in Poperinge, Belgium, estimated that 10,000 pickers came from nearby and 10,000 from outside the immediate area.

During Hoppefeesten, the festival held every three years in Poperinge, volunteers demonstrate how hops were picked relatively recently in one of Belgium's best known hop growing regions.

Even as those memories become distant, they were once such an important part of the local culture that they are preserved. The hop museum in Poperinge has an oral history, recorded only in 2011, from Bertin Deneire, a teacher and museum guide. "For us kids, hop picking was a frightfully tedious job. Like the adults, we tried to break the monotony of the dreadfully drab days in the field doing the same job for hours on end. Kids would catch ladybirds in matchboxes, tell silly riddles, or play pranks on each other, whereas the adults swapped gossip, had singsongs, told jokes (which I didn't always understand), and had hilarious farting competitions," Deneire remembered.

Not surprisingly, some of his strongest memories were tied to aroma. "I can smell the scent of drying hops now, an indescribable, fragrant smell that I can only define as 'somewhere between roast apple and blooming geranium.' For whoever went hop picking it became a scent that could never be erased from their memory, much like remembering an old addiction," he said.

His final thoughts, not surprisingly, were of the last day of harvest, which concluded with a dance and a final trip to the hop yard: "Goggle-eyed I used to stare at this incredible spectacle. Elderly people whirling like dervishes, falling over after a wild waltz. Men slapping young girls' bottoms, fat women sweating profusely. The boisterous atmosphere in the farmhouse became sultry with the perspiration of this 'madding crowd,' and soon we would leave the room and follow the foreman into the dark of night. Singing and in droves we went back to the field to witness the hop guy (a straw dummy) being 'hanged' from the last pole of the field and set alight with a burning torch made from a rolled-up newspaper. In a way it looked like an execution, our revenge against nature for weeks of pent-up frustration and endless toil under the burning sun. And we would dance and sing around the guy until the last embers of the bonfire had died out."[1]

The museum still keeps a straw hop devil on display. Farmers feared the gusty winds and thunderstorms that could destroy a hop yard in the final days before harvest and hung the large straw figures in their fields to ward off bad weather.

Hop harvest season was romanticized in books, song, and images, such as paintings or photo postcards of pickers in the fields. A drawing in *Hop Culture in the United States*, published in 1883, shows women sitting in what appears to be a well kept parlor with hops hanging like a large bunch of grapes. They are picking hops at their leisure, a baby playing in the middle of the room, another child sitting on a woman's lap, and a dog sleeping off to the side. An English writer in the magazine *Land and Water* suggested, "There is something wonderfully soothing in the aroma of the hop. The pickers sleep well in the little huts they have run up; and the babies have been swinging in their hammocks between the tall bines and open network of strobiles, invariably in the arms of Morpheus."[2]

Harper's magazine painted an equally engaging picture in 1885: "Hop-pickers expect to live on the fat of the farmer's land, and, as a rule, they are not disappointed. Whole sheep and beeves vanish like manna before the Israelites in the short three weeks that follow, while gallons of coffee, firkins of butter, barrels of flour, and sugar by the hundredweight are swallowed up in the capacious maw of the small army. ... The hop dance is an indispensable adjunct to the picking season, much counted on by the gay throng but a good deal frowned upon by the staid and proper seniors. Like many other creations which have had their origin in a harmless beginning, it has often run away with propriety and brought scandal in lieu of innocent pleasure."[3] These harvest dances later gave a name to school dances (for instance, a sock hop).

George Orwell questioned this romanticized view, attacking upfront the idea it was a "holiday with pay," experiencing firsthand that it was nearly impossible to earn as much as advertised by traveling to Kent and working as a picker. In an essay in *New Statesman & Nation* he allowed, "There is no pleasanter place than the shady lanes of hops, with their bitter scent—an unutterably refreshing scent, like a wind blowing from oceans of cool beer. It would be almost ideal if one could only earn a living at it." Although government officials inspected workers' accommodations, Orwell was not impressed. "But what it can have been like in the old days is hard to imagine, for even now the ordinary hop-picker's hut is worse than a stable," he wrote in 1931. "My friend and I, with two others, slept in a tin hut ten feet across, with two unglazed windows and half a dozen other apertures to let in the wind and rain, and no furniture save a heap of straw; the latrine was two hundred yards away, and the water tap the same distance. Some of these huts had to be shared by eight men—but that, at any rate, mitigated the cold, which can be bitter on September nights when one has no bedding but a disused sack."[4]

The Society for Promoting Christian Knowledge described conditions far worse in the 1860s, although they in turn were considerably improved over those in decades before. Author Rev. J.Y. Stratton compared lodgings to "dog-hole" dens and hovels in *Hops and Hop-Pickers,* sparing neither hop yard owners nor immigrant workers. He wrote that it was difficult for the Society for the Employment and Improved Lodging of Hop-pickers to carry out its plans to upgrade housing because of the "lawless and predatory habits of those who it desired to benefit." Stratton wrote, "Half monkey, half tiger, the typical hopper was often a thief."[5]

Conditions weren't necessarily better almost a century later. One contributor to *Hertfordshire: Within Living Memory,* a collection of oral histories documenting life in the West Midlands during the first half of the twentieth century, recalled that the workers who came from the depressed Welsh mining valleys in the 1930s were often housed in very primitive conditions, sometimes in pigsties. However "this didn't seem to worry them in the least so long as they could earn a few bob, so that

Father could spend Saturday night in the pub and Mum could buy the kids some shoes."[6]

"Home pickers" and many workers who returned annually enjoyed better accommodations. "In September and October (fulltime) workers at Cradley were greatly increased by pickers who came from the Black Country and Wales for their summer holidays and to earn some extra money," a contributor said. "The same families tended to go back to the farms year after year. They were joined by gypsies, and the local school children also had their holidays then. They all lived in barns and mobile homes especially kept for them."[7]

When hop growing turned into an industry in the United States in the nineteenth century, farmers had to recruit extra workers. J.D. Grant in Sonoma County, California, did it by appealing to campers. His advertisements described the natural beauty of the Russian River area. He offered "wood, water, and pasturage for horses, as well as employment for picking hops." J.F. Clark in Otsego County, New York, put special cars on a train that traveled from Albany with hundreds of pickers, locking them in lest they be tempted to get off at a stop along the way.

According to the *Otsego Farmer*, "Hop picking may not be as fashionable as golf but it has its compensations; the money comes handy, there is something about the season that is like a call for a gathering of the clans of old, and those who are ashamed of it when they get back to town can get the stain off their fingers with time and Hand Sapolio (a soap)."[8]

On the West Coast, magazines portrayed the Washington hop growing region as a tourist attraction. "There is a wonderful charm about a large Hopfield in the harvest season; everywhere is the perfect symmetry of Old Greek architecture," according to an article in *Overland Monthly*. "There are endless vistas, cool, green, and inviting, stretching away between the hop poles."

The writer also described colorful Indian encampments. The Indians called harvest "hops time" and in Washington did most of the picking. About 2,500 traveled to Puyallup from parts of the Puget Sound, Alaska, and British Columbia in 1882. The majority arrived by canoe and the largest of those carried 20. The most experienced pickers earned $3 a day; $1.25 was average.[9]

Members of the Sinkiuse, Wanapum, Nez Perce, Yakima, and Okanagan tribes all picked hops in the Yakima Valley. According to a history of the Moxee region, "On weekends, the Indians gambled and many people went to see them play the 'Stick Game,' the 'Bone Game,' dice, and cards. ... At the end of the hop season the members of other tribes would congregate for a final big game and would bet their horses, saddles, blankets, baskets, or anything else they had. This would go on all day Saturday and Sunday. Some even gambled their entire earnings of the season."[10]

Orwell might have wondered about the economic sense of hop picking, but he couldn't deny its appeal. "Yet the curious thing is that there is no lack of pickers, and what is more, the same people return to the hop-fields year after year," he wrote. "What keeps the business going is probably the fact that the Cockneys rather enjoy the trip to the country, in spite of the bad pay and in spite of the discomfort. When the season is over the pickers are heartily glad—glad to be back in London, where you do not have to sleep on straw, and you can put a penny in the gas instead of hunting for firewood, and Woolworth's is round the corner—but still, hop-picking is in the category of things that are great fun when they are over. It figures in the pickers' mind as a holiday, though they are working hard all the time and out of pocket at the end."[11]

Turning Acres of Hops Into Bales

When Jason Perrault's great-grandfather began farming hops in the 1920s, it took 100 people 30 days to harvest 13 acres. When his grandfather ran the farm, it took 80 people 30 days to harvest 150 acres. Today 40 people harvest 750 acres in 30 days.[12]

Everything at Otto Scheuerlein's farm in Spalt happens on a smaller scale, but with their two sons—who take holiday to help—the Scheuerleins harvest about 10 acres of hops in nine days. Farms in Yakima, or even other parts of Germany, have machines dedicated to quickly cutting acres of bines from their trellises, large stationary picking machines, and giant kilns. The Scheuerleins have an attachment on their tractor, a small picker, a three-drawer kiln, and a conditioning floor on the second story of the barn, with a picker and baler on the first floor.

The picker strips the cones from the bines and leaves before a series of conveyor belts and fans separate the leaves from the cones. On a large farm it is a very noisy process. Stepping away from all of that Perrault motioned toward where the hops would dry. "I love when you first lay the kiln and turn on the heat. The greenness of the aroma. And it's so calm," he said, before turning back to the clatter. "It's an interesting dichotomy, the violence of the picking, the quiet of the kiln."

Curiously, the violence is less a threat to what brewers want from the hop than the calm. Val Peacock recognized that more than 20 years ago when he first began visiting Anheuser-Busch's field offices. "It was pretty obvious to me that (drying) is the most critical thing a farmer does," he said.

Brewing manuals from the late nineteenth century described different oast house configurations in detail and offered suggestions, which often differed, on proper temperatures for drying. In 1891 Herbert Myrick wrote, "It is claimed the 'natural cure' preserves far more of the essential oils and other brewing principles than is possible by the artificial hot-air cure in England and America, and that this accounts in part for the peculiarities of Spalt hops that command such extraordinary prices."[13] Although Spalt farmers no longer dry hops in lofts, those buildings with high, steep roofs still loom over parts of town.

Modern-day American and European growers dry hops much differently. Kilns in the U.S. Northwest consist of multiple sections within what are basically giant sheds. Conveyor belts carry the cones to the kilns after they are separated from the bines, and they are spread 24 inches to 36 inches deep. Heated air, forced through the bed from the bottom, dries the hops within six to eight hours.

Farmers choose to spread different varieties to different heights and dry them at different temperatures. Double R Ranch dries aroma hops between 130° F (54° C) and 140° F (60° C), with Cascade at the higher end of the range because its cone is durable. The ranch dries high alpha hops at 145° F (63° C). John Segal said the Segal Ranch dries its Cascades at lower temperatures (which he chooses to keep proprietary) than at any other farm, resulting in a process that takes two hours longer. "We're drying for aroma," he said. In 2010 Segal's Cascades contained 3.1 percent essential oils, almost double typical Cascades. "We let them hang longer (before picking), and we dry for aroma," he said.

Segal employs a unique way to make sure the hops dry evenly. To illustrate he put on a pair of gloves that hung beside a rope on the side of a kiln. He tugged the rope, and the full length of it emerged from the hops, reaching to the other side of the kiln. Before workers lay a bed of hops to dry, they place these ropes across the floor. Yanking them to the surface effectively stirs the hops. "No wet spots, no flavor blowouts," he said.

At Perrault Farms, workers limit the depth of Simcoe and Citra, two hops high in alpha but mostly valued for their aroma, to between 24 and 28 inches and dry them at about 130° F (54° C). Jim Boyd, hop processing and logistics manager at Roy Farms, said the company estimates that drying all its aroma hops at 125° F (52° C) takes 20 percent longer and uses 18 percent more diesel fuel.

Farmers have many tools to monitor the moisture within the kiln, realizing it will never be uniform but hesitating to invade it too many times because that means breaking more cones. "I would tell you, the best results are from an experienced drier who can (assess an entire kiln) by picking up a few cones," Kevin Riel of Double R said. "I think this is the most artistic aspect of controlling quality. There are very few people in the valley, and they are much sought after."

Workers transfer the dried hops to cooling bins for 24 to 36 hours for homogenization before they are baled.

Most commonly in Germany, although certainly not on every farm, growers dry hops in three-level kilns, then condition them before baling. What appear to be drawers, particularly at smaller farms such as in Spalt, have louvered floors so hops can be dropped from one layer to the next as drying proceeds. Fresh hops are loaded onto the top tier each time that dry hops are removed from the bottom tier. As in the United States, they target 9 to 10 percent moisture in the bales.

Belt or continuous driers are a variation on the three-level system and are found more often in the Czech Republic. Hops pass across the heat source on moving belts, beginning at the top level, falling to the next, reversing direction, falling again, and reversing again. The belt drier requires less labor.

Rubbing and Sniffing

Roy Farms produces its own pellets, milling cones right after they are dried. Most often, farmers ship baled hops to processing plants, where workers break up the bales, mill them, and make pellets. The pellets amount to a small part of Roy's production, and, obviously, the company can make its Harvest Fresh Pellets only during harvest season. "It's much more ecologically friendly," said Leslie Roy. "We feel there is some benefit in the preservation of oils, and we have brewers who feel that way, too."

Larry Sidor spent seven years as general manager at S.S. Steiner, where among other things he helped design and install nitrogen-cooled hop pellet dies. He appreciates what makes Harvest Fresh Pellets different. However, as an advocate for brewing with whole cones and a veteran of selecting those hops each year, he has reason to pause.

"From a pelletizing standpoint, it is a dream come true. From a brewer's standpoint, what are you going to do, select them going into the pelletizer? Camp out?" Sidor asked, rhetorically. "There are brewers who select from pellets, but I can't say I've mastered that art. They are not as revealing of what I'm looking for."

After selecting hops for more than 30 harvest seasons, first at Olympia Brewing and then at Deschutes Brewery, Sidor has a good idea of what that is. Before he left Deschutes at the beginning of 2012 to start his own brewery, Sidor would rotate only one new member into the five-person selection team at a time. The first thing she or he discovered was "best hop" wasn't necessarily the best choice.

"They learned we weren't selecting to personal preference. There were hops it pained me to leave behind," he said. "But I knew they weren't going to deliver the compounds we want to the beer. We had to select the profile we selected in the past." Deschutes reassessed its choices quarterly, something easier to do because the brewery uses flowers almost exclusively. "We'd rub them, run the storage index, check the alphas and betas, see if we really got what we thought we were getting," he said.

Most small breweries cannot devote the resources to visit the Northwest or another hop growing region for harvest. In addition it would be impossible for hop suppliers to accommodate them all. However, brewers may use advice from experienced buyers (such as John Harris on the pages below) when dealing with their hop brokers.

"It's like a chef going to the fish market early," New Glarus Brewing co-founder Dan Carey said, explaining why he regularly travels to both Europe and the Pacific Northwest to select hops. "You have to remember you are buying farm products."

That's true on the largest scale. "We've always selected hops on a sensory basis," said Steve Dresler, brewmaster at Sierra Nevada Brewing. "I don't look at alpha. I don't look at oil. We open up the brewers' cuts and get dirty," he said. Founder Ken Grossman, Dresler, and those on the brewery's selection team will choose hundreds of thousands of pounds of hops in 36 hours. "We're aroma buyers. I'll do the same for alpha through the mail."

Dresler appreciates the annual trip for more than the hops he buys. "You always get to see new stuff, chatting with people," he said. "It's the benefit of sitting down and having beers together."

Victory Brewing co-founder Ron Barchet travels to Germany every year to select hops. "I like to work with farmers," he said. "We're not going to get the same hops if we're not directly involved."

Hop farmers in Tettnang grin in recognition when they hear his name. "Ron has a very fine hop nose," said Georg Bentele, a farmer Victory has long-term contracts with. "He can differentiate, he can taste the (different varieties) in the beer. That is not so easy."

Barchet isn't looking for the aroma he will find in a beer when he rubs and sniffs hops in Tettnang. "(The aromas) are almost unrelated. You have to know what you are smelling and know what that will translate to in the beer," he said. "I'm looking for freshness."

He has learned from experience. "Sniffing them. Using them for a year. Sniff them the next year. I don't know of a better way," he said. "Teas really don't work. You don't know what you have until you ferment it."

Whole cones from unbaled lots, processed into harvest fresh pellets, are displayed with brewers' cuts as they await evaluation and selection.

A Brewer's Guide to Evaluating and Selecting Hops

John Harris worked as a brewmaster for Full Sail Brewing, Deschutes Brewery, and McMenamins before starting his own brewery in 2012. He received the Brewers Association Russell Schehrer Award for Innovation in 2001 and has been evaluating and selecting hops for more than 20 years. Brewers began using his guide to hop selection after a presentation at the 1999 Master Brewers Association of the Americas convention. He updated it for this book.

Common Flaws in Hops

When you find a flaw, it is important that you look at it in relation to the whole sample. For example, finding a few windburned cones in the brewers' cut is not that uncommon.

Spider mites. These small insects love hot weather and dust, which means they normally appear in late summer, just as the cones are ripening. They kill the cones, and dead cones have a reddish tinge.

Hop aphids. Aphids like cool weather and so are more predominant in the spring. They burrow into the hop, suck the life out of the cone, leave their waste, and die. What's left is a black, moldy mess that leaves the cone unusable.

Mildews. Two types of mildews affect hops: powdery and downy (peronospora). They both can affect the hop at any stage of development but are most damaging early on. Downy mildew is related to warm, wet weather and is environmentally driven. Oregon has downy mildew, while Yakima rarely does, and Idaho has none at all. Powdery mildew is a systemic infection that is born out of the soil and lives with the hop year-round. It was found in Yakima starting in 1997. Both kinds keep the hop cones from developing properly by totally stunting their growth. Of the two mildews, powdery affects the grower the most. Hops that have been exposed to either mildew near harvest time may have some silver or brown spots but may still be usable. That is up to the brewer. You probably will never see any hop samples that have been severely damaged. Only in a bad year might you see the diseases in your samples.

Windburn and spray burn. Evidence of windburn and spray burn shows on the cones as brown discoloration. When the spray is applied, the grower uses a large fan that blows the spray over the field at about 130 miles per hour. This can cause stress to the hop cones and burn them, either by the chemical or wind force. High wind conditions may cause the cones to bump into each other, causing bruising to the cones. This is more of a cosmetic flaw and should not hurt brewing performance. Evaluate the hops to determine if they meet your needs.

Hop Selection Team

When evaluating beer flavors, you rely on a set framework for tasting procedures, a lexicon to describe what you taste, and an educated taste panel to help you make decisions. When selecting your hops, you should also have a trained and consistent team to make the decision for your company. The selection team should be between four and six people. This size allows all the members to have input in the decision. You need to train your group, so they understand the aroma lexicon you are going to use. You all need to be on the same page and speak the same language about what you smell, feel, and see in the hops. I think it is important to make sure that the people who are selecting the hops also have an understanding of what effect the selection will have on the finished beer and see the hops in use in the brewery.

The team needs to evaluate the samples of a given variety and judge them against the same variety. This is something worth repeating—remember the importance of the variety in the evaluation process.

Varieties differ in appearance, aroma, color, and many other variables to the point where they cannot be judged against each other. They can only be judged based upon what they will contribute to your beer. Even within the same variety, you will see big differences in aroma and appearance, depending on what country, state, or field they are from. Sometimes the duller sample is the best one for your beer. As brewers our challenge is to make a consistent beer time and time again.

One year we were looking at Cascade hops and found one lot that was just spectacular. It had the trademark Cascade stink of fresh citrus and slight fruit. It was clearly the most aromatic of all the lots we rubbed, but we did not select it. Why not? Because after years of rubbing Cascades, this was one of the most intense samples we had ever seen. Our fear was that it would be too difficult to find a sample like it in future years. Instead we selected a lot that we thought was just as nice and would keep our beers' flavor more in line with where we wanted it to be year to year. Each year's growing conditions will affect what hop choices you have. Some years you see high alphas and sometimes low. When looking at samples, have set goals for what you're looking for and a way to communicate what you're picking up in the hops. Sometimes a sample will provide the aroma you're looking for but not the analytical parameters you have set for the particular variety. Remember to evaluate by asking the question, "Is the look, aroma, and feel right for this variety?"

Just like at a tasting panel, the team should refrain from wearing perfumes and should have clean but not soapy hands, and the selection should take place in a properly lit, comfortable environment away from noise and any odors.

The Brewers' Cut
A brewers' cut is a sample of hops removed from every fiftieth bale from a field. The samples are analyzed for alpha, beta, and hop storage index (HSI). The brewers' cut is divided into two brewers' samples. When evaluating, you are given a sample from each lot you are considering.

In addition to the analytical data, you are also given a leaf/stem (L/S) and seed count. The U.S. Department of Agriculture does these on every tenth bale in the lot. The percentage given is based on weight. Scores from 0 to 1 percent are given 0 as a score, 1 to 2 percent is given a 1, and 2 to 3 percent is called a 2. This score is important because it indicates hop product quality. High LS/S numbers mean less brewing material and possibly dirtier hops. It is all a reflection of how the grower processed the hops. It is important to remember that some hop varieties are prone to having more seeds. Also, the growing conditions and locale may affect seed production.

Hop Rubbing Descriptors
It is good to use common words to describe what you smell in hops. Here is a list of positive and defect descriptors used in the hop industry.

Positive. Forest-woody, mint, citrus, piney, spicy, grapefruit, estery, grassy (fresh), resinous, floral, herbal, cedar, stone fruit (peaches, plums), tropical fruit (mango), tomato plant (not tomato per se).

Defects and off-aromas. Earthy, grassy (brown or dead), musty, kerosene, haylike, strawlike, tea, cheesy, onion, garlic, sweaty, tobacco.

Hand Evaluation of Hops

Remember to evaluate by asking the question, "Is the look, aroma, and feel right for this variety?" Your evaluation should start with lower alpha/aroma hop varieties and work up to high alphas. It is also good to take short breaks between varieties to clear your head and nose. Ask your supplier for general information on what you should be looking for in the samples. If you know that a certain flaw is common in the hops in a particular year, you may be able to move through your samples faster. Ask the suppliers how they evaluate hops and what they consider important.

1. Examine the sides of the brewers' cut. A lot can be learned from looking at the sides. The sample has three cut sides and one uncut side. First, look at the uncut side to see what the hops look like on the inside of the bale. The density of the bale will play a role; German bales are packed more loosely, so that a cut is not necessary for assessment, as the cones don't stick together very tightly. Looking at cones that have not been cut, you can check the cone stability and shatter, which is when the petals just fall off the strig due to low moisture. Look at the cut sides. How does the lupulin look? It should be a nice, bright yellow or light orange. If it is a deep orange color, it may be a sign of oxidation, too much heat in the kiln. Rub the cut side to feel for any seeds. Do you see any leaf or stem?

2. Feel for moisture. Take your hand and press down on the hop sample. How does it feel? It should have a nice firmness with a slight spring when released. This indicates that the hops were properly dried and baled. The harder the sample feels, the more moisture it has. If the sample is too moist, it may go through a secondary sweat, causing raggy cones like a wet rag. If the cones are wet and raggy, they will not fall apart when being rubbed. When the sample is hard, it is referred to as boardy, like pressing on a board. If the sample is too dry, it will shatter when pressed, and the hops will seem lifeless. This can be an indication of too-late harvesting. Lower alpha/aroma hops tend to break apart more easily than high alpha hops due to less lupulin and essential oils.

3. Inspect whole cones. Take the cardboard divider and cut off a 2-inch sample. Break the cones apart and inspect for flaws. Look for wind or spray burn, aphid or spider damage, mold, leaf, and stem. Check the cones' sizing—are they the right size for the variety? Also check the strig in the cone. Are the hop petals attached well? This is a sign of proper drying. The whole cones should outnumber the broken by a great majority. The more broken cones you have, the greater chance of oxidized hops. Break open a cone and inspect the lupulin glands. How do they look? Is the color right for the variety?

4. Assess the hop color. Does the general hop color look right? Are they green, yellow, or brown? If the sample is discolored, it may be due to too-late harvesting, abuse in the kiln, insects, or hot weather. In general, Oregon hops are not as bright as Washington hops. What is the sheen of the hops? Are they dull, pale, bright, or brilliant? You will find that the colors vary. Only experience will teach you what is normal. Look at hops in respect to variety and growing area. The environmental growing conditions can also vary from year to year, which will affect the color. Looks alone are not a reason to reject or accept a hop lot.

5. Whole cone aroma evaluation. Take a sample of unbroken cones and evaluate their aroma. In unbroken form you should be able to detect any off-aromas. Review the list of aroma defects.

6. First rub—the light one. Take a sample, lightly rub it in your hands, and set aside. This helps rinse your hands to move properly between samples. Take another sample, and lightly rub it in your hands, making sure that you start to break apart the lupulin glands. Take a smell. How does it smell? The light rub is a good way to look for any grassy notes.

7. Big rub—release the aroma. Take your lightly rubbed sample and crush it in your hands. The hops should fall apart. This rub releases the hydrocarbons in the oils and lupulin glands. Feel the sample for moisture. How does it feel? Do you feel the hop oils? High alpha hops will be stickier than low alphas. Give the sample a big smell. Evaluate it using the known hop descriptors and any other aromas you may be picking up. Do you like it? Is it true to type? How will this hop sample work in your brewery?

8. Big rub—hold that sample. Keep the big rub sample in your hand for a minute to warm it. Give it another smell. Does it smell the same? Do you still like it, or do you like it better?

9. Discuss the rub. Look at all the samples of a given variety before selecting with the team. Rub the samples, take notes, and then discuss. Are the hops true to type? How will they work in the brewery? Will they provide repeatability in brewing? Which one do you like best? Why?

10. Choose your lots. Check to make sure your choice of hops supplies the alpha, beta, oils that you're looking for along with good HSI values.

It is good to have a rating scale to work with when you do your selection. Use the framework of the selection process to come up with an evaluation form. Each variety also has trademark characteristics that you may want to look for when selecting. Assign a score to each sample and rate per your criteria.

Further Evaluation

After selecting your hops, use your lab and brewery to evaluate their performance.

Hop teas. Hop teas can be used to check the hop aroma of either leaf or pellet samples. Many times, the initial aromas of teas will have grassy notes due to the chlorophyll or strange aromas. I make it by taking a liter of water, adding a handful of cones, and bringing it to a boil. Allow the tea to simmer for a while to drive off the myrcene and other volatile compounds, since you would never find these in your beer after proper boiling. Smell your tea at different times along the way, to see how additional time in the kettle may affect the aromas in the beer. It may also be sufficient to mix hot (not boiling) water with a defined amount of hops for 10 to 15 minutes to mimic a whirlpool addition. An extraction with cold water may give you a hint of how the hops perform if used for dry hopping. If you're doing hop teas, it is important to develop a baseline to compare one lot to another. Sometimes the aromas you get out of hop teas are not that pleasant.

Evaluate in plant. Run brewing trials to see how well the hop performs in your beers. To truly evaluate the hop it must be put into a malt base and fermented. Start with a lighter house beer or brew a pale malt-only brew and hop it around 12-20 IBU, so that you can really see the character of the hop variety. For an aroma variety you want to make sure you use enough to obtain a sufficient amount of hop oil. Evaluate with your taste panel. Once you find a hop you like, see how it works in your regular beers.

Evaluating Pellets

The best pellets start with high quality hops. Your pellets will only be as good as the hops used to produce them. Know your suppliers' process.

1. Warm sample to room temperature. This will allow the aroma to be released. If the sample is too cold, the aroma is locked in, just like in a beer.

2. Examine appearance. The pellets should be green in color, but it will vary depending on variety. Dark olive and brown pellets indicate the possibility of oxidation. Keep in mind that the incoming hop color will affect the color of the pellet. A glassy appearance is a sign of excess heat during processing.

3. Finger smash. Rub the pellet between your fingers; with a little effort the pellet should be able to be broken down with your fingers.

4. Evaluate the aroma. The pellet should have a fresh hop aroma. Check for cheesy aromas and other signs of oxidation. Evaluate with a hop tea if you like doing teas.

A Checklist

Selecting hops is a personal thing. Each brewer and brewery will approach it from different angles. Some key points to remember:

• Learn to identify the flaws in hops and how they can affect your beer.

• Get to know the aroma, feel, and appearance of your favorite varieties.

• Develop your aroma vocabulary. Learn the common hop descriptors. Tune your sense of smell as well as your tasting palate.

• Establish a team to help with selection.

• Create process for evaluation. Establish guidelines and follow them for each sample. Create a personal evaluation form.

• Select for consistency.

• Develop a strong relationship with your suppliers.

Notes

1. Bertin Deneire, "The Hoppiest Days of My Life," oral history, kept at HopMuseum in Poperinge, Belgium. Available at *www.hopmuseum.be/images/filelib/hopstory.pdf.*

2. P.L. Simmonds, *Hops: Their Cultivation, Commerce, and Uses in Various Countries* (London: E. & F.N. Spon., 1877), 79.

3. G. Pomeroy Keese, "A Glass of Beer," *Harper's New Monthly Magazine* 425 (October 1885), 668.

4. George Orwell, "Hop-picking," *New Statesman and Nation,* Oct. 17, 1931.

5. Rev. J.Y. Stratton, *Hops and Hop-Pickers* (London: Society for Promoting Christian Knowledge, 1883), 54.

6. Hertfordshire Federation of Women's Institutes, *Hertfordshire Within Living Memory* (Newbury, Berkshire, England: Countryside Books, 1993), 163.

7. Ibid., 164.

8. L. Gimble, R. Romanko, B. Schwartz, and H. Eisman, *Steiner's Guide to American Hops* (Printed in United States: S.S. Steiner, 1973), 55.

9. Gimble, et al., 56.

10. Alice Toupin, *MOOK-SEE, MOXIE, MOXEE: The Enchanting Moxee Valley, Its History and Development* (1970), 8.

11. Orwell.

12. Here's the math in more detail. 1920s: 13 acres, 100 people, 30 days = 2.3 acres per day, 7.7 people per acre; 1960s: 150 acres, 80 people, 30 days = 5 acres per day, 0.5 people per acre; Today: 750 acres, 40 people, 30 days = 25 acres per day, 0.05 people per acre.

13. Herbert Myrick, *The Hop: Its Culture and Cure, Marketing and Manufacture* (Springfield, Mass: Orange Judd Co., 1899), 177.

Hop yards in the Tettnang region of Germany. Those are the Alps in the distance.

Low-trellis hops grow in an experimental field in the Czech Republic, with high-trellis hops in the background. See page 96.

This oast house in Sussex is one of about 3,500 remaining in England. Most have been turned into homes. (Photo courtesy of Donar Reiskoffer)

The Hop Kiln Winery near Healdsburg is one of the last hop kilns left in California, a reminder of when Sonoma County produced a significant amount of hops.

2011 Hallertau Hopfenkönigin *Veronika Springer (center) and her court. Christina Thalmaier, the 2010-2011 queen, stands to her left.*

The highlight of Hoppefeesten, *a hop and beer festival held every three years in Poperinge, Belgium, is a parade that snakes through the town. On this float, children dressed as hops sit atop trellises.*

The germplasm collection at Czech Hop Research Institute in Žatec contains long-held varieties that can be cultivated for use today.

A 40-meter-tall "hop lighthouse" overlooks a 4,000-square-meter hop museum in the center of Žatec.

Peter Darby in the Wye Hops Limited experimental yards at China Farm near Canterbury.

Breeding bags isolate plants at Wye Hops after they have been cross-pollinated (so they will not be naturally pollinated again).

A walkway off the second level of the HopfenMuseum outside of Tettnang allows visitors to stroll at eye level with the tops of hop plants.

At the Deutsches Hopfenmuseum in Wolnzach, this five-meter-tall cone dispenses odors of various hops.

Although Belgium grows a relatively small amount of hops today, how important the plant was to Poperinge is apparent on a daily basis. Above, stained glass windows in a church at the city center include scenes of hop pickers at work. Below, a giant cone and trellises decorate a roundabout. The town also has many streets named after other hop growing regions, such as Saaz.

Sierra Nevada Brewing has had to commission many new "torpedoes" as Torpedo Extra IPA *has grown in popularity. They can link together to dry hop large tanks.*

Bales of hops must be broken up before they can be used at Sierra Nevada, which brews with only "whole hops."

6

The Hop Store

A variety of varieties come in a variety of forms

Just a few days before he left Deschutes Brewery to start his own brewery at the beginning of 2012, Larry Sidor repeated one of his favorite stories. He talked about what it took to convert Olympia Brewing from a facility that used whole hop cones to one that brewed with pellets. "It is my biggest regret in life," Sidor said. "You get one do-over in life. Deschutes was my do-over."

Sidor worked seven years at S.S. Steiner in the Yakima Valley after leaving Olympia and before joining Deschutes. He saw significant advances in hop pellet processing. He can recite the positive attributes pellets offer. They provide higher levels of utilization, particularly when used in dry hopping, and more consistent amounts of alpha acids. They take up less room for storage and store better. He'll acknowledge that differences in aroma and flavor that result from brewing with pellets rather than whole cones may be subjective, but if such things are a matter of opinion he has his own.

Fuller's Brewery in London converted from cones to pellets in 1970. "The first three months (after harvest) whole hops may be better," said John Keeling, brewery director. "The next three they are rather even, but the following six months pellets are much better."

It was only in 2008 that Anheuser-Busch converted from using whole hops to pellets, making Sierra Nevada Brewing the largest cone-only brewery in America.

"We're well aware of the pros and cons of pellets," said Sierra Nevada founder Ken Grossman. In the 1980s the brewery split a batch of *Celebration Ale,* dry hopping one portion with pellets and the other with cones. Grossman said everybody preferred the batch with whole hops. "Philosophically, we're committed to flowers," he said.

Sierra Nevada is not alone. Victory Brewing in Pennsylvania uses only whole cones. "We don't think (pellets) are as good," co-founder Ron Barchet said. "A few more brewers are starting to see the light." Vinnie Cilurzo at Russian River Brewing makes most of his beers with pellets—even using a "hop cannon" to fire them in during dry hopping—but said, "I find a softer, more subtle hop aroma from flowers."

Those differences aside, most discussions about "traditional hops" include cones and pellets in one category and everything else in another.

Hop extracts are not exactly new. The New York Hop Extract Company built the first extraction plant in the world large enough to produce quantities sufficient to supply brewers in 1870. J.R. Whiting licensed the process in 1875 and erected a factory in Waterville, New York, at the center of the state's hop growing region, twice enlarging it to meet demand. Brewers bought extract when hops were plentiful and prices low as insurance against poor crop years and high prices, using it in combination with whole hops. The Waterville factory closed after the turn of the century and the New York City plant in 1933.

In 1963 the Hops Extract Corporation of America built a large plant in Yakima. By 1971 the equivalent of 23 million pounds out of the U.S. crop of 49.6 million pounds were converted into extract.

Pelletizing and Pellet Products

Sidor suggests brewers "should be there when they pelletize your hops." Few will find that practical but may benefit from seeing hops processed at least once and understanding how it is done. Hops are first milled in a hammer mill, preferably under refrigeration to reduce losses of alpha acids and essential oils. Hop powder is accumulated in a mixing vessel and then pelletized. Pelletizing mills have either a ring die or flat die configuration. A ring die machine introduces hop powder into the interior of a rotating cylinder (the die), allowing rollers to press the powder though the holes in the cylinder. A flat die machine utilizes a horizontal die on which hop powder is distributed while rotating rollers force the powder though holes in the die. High quality pellets result from low pelletizing temperatures, and the lower density pellets generate lower temperatures.

Almost all the lupulin glands are ruptured during pelletizing, one result being the brewing utilization rate of hop pellets is 10 to 15 percent higher than that of whole hops; another that, unprotected, they oxidize three to five times faster than cones. Pellets may also be better homogenized, since processors blend them during pelletizing, resulting in more consistent levels of alpha acids and essential oils.

Type 90 pellets. T90s once contained 90 percent of the nonresinous components found in hop cones, thus the name, although today product losses are generally less and the percentage is actually higher. They are the most common sold. The composition of oils and alpha within the pellets is similar to cones but not necessarily identical.

Type 45 pellets. T45, or lupulin-enriched, pellets, are manufactured from enriched hop powder. Processors mill the hops at about -20° F (-30° C), which reduces the stickiness of the resin, and separate the lupulin from unwanted fibrous vegetative matter. Although the name implies the hops are enriched to twice the level of T90s, the level may be restricted by the amount of lupulin in the original hop. Normally, processors customize the level, producing, for example, a T33 or a T72 pellet that would still be referred to as a T45 pellet. T45 pellets are produced more often from low alpha aroma hops.

Isomerized pellets. Referred to as isopellets and also referred to as pre-isomerized, meaning the conversion of alpha acids to iso-alpha acids occurs during production. That is achieved by adding food-grade magnesium hydroxide and treatment under heat. They provide higher utilization and need only 10 to 15 minutes of contact time with wort to achieve that.

Type 100 pellets. Also known as hop plugs, these are compressed whole cones used primarily for dry hopping cask-conditioned ales.

Hop Extracts

Scientists used water and ethanol to extract hops in the nineteenth century. Today most processors employ carbon dioxide extraction, either supercritical (Europe and the United States) or liquid (England). At its simplest, hop extraction involves passing a solvent through a packed column of pellets to collect the resin components and removal of the solvent to yield nearly pure resin extract.

Liquid CO_2 extraction produces the most pure whole resin and oil extract, including none of the hard resins or tannins and lower levels of plant waxes, no plant pigments and less water and water soluble materials. The supercritical process, used for 95 percent of extract produced, yields slightly more alpha acids, usually between 35 and 50 percent, with hop oil content between 3 and 12 percent. As with pellets extracts may be isomerized during processing and have a much higher utilization rate.

The cost per bitterness unit may appear higher with extracts than with whole hops or pellets, but its advantages—reduced costs for shipping and storage, uniformity, stability, better utilization, and reduced wort losses—offset that. Although many people associated with craft beer consider extract the antithesis of natural hops, a product used primarily in beers with little or no hop presence, it also may enhance the quality of highly hopped beers.

Cilurzo was the first American craft brewer to speak openly about using CO_2 extract. Initially he brewed *Pliny the Elder*, a groundbreaking double IPA, using only hop pellets. He didn't like grassy, chlorophyll flavors he attributed to sheer hop mass. Following a suggestion from Gerard Lemmens at Yakima Chief, he replaced pellets with extract for the bittering addition.

"We kept it secret for the first few years," Cilurzo said, "but Gerard twisted my arm." Cilurzo gave Lemmens permission to publish the information in a Yakima Chief newsletter. Scores of other small breweries soon began to use CO_2 extract.

Jeremy Marshall said that in 2010 Lagunitas Brewing in California started using extract in a wider range of its beers, because the brewery wanted to lower the level of tannins in several brands, an issue he attributed primarily to a poor malt crop. "We use hop extract because it increases the quality of our beer," he said.

Cilurzo includes varietal extracts in several of his beers, such as Amarillo midboil in both *Blind Pig IPA* and *Pliny the Younger*. "Now they'll extract anything for you," he said. He visits European breweries often enough that he understands even relatively small breweries there are more likely than U.S. craft breweries to use products labeled "advanced hop products" by advocates and "downstream products" by detractors. Those do not interest Cilurzo, so he understands why many brewers question using even CO_2 extract.

"It's such a personal decision," he said. "It's philosophical."

Advanced Hop Products

Advanced, or refined, hop products are extracts that have been created for a specific purpose.

Rho hop extracts. Rho extracts may contain only dihydro-iso-alpha-acids (rho) plus beta acids and hop oils, or only rho. They help prevent light-struck flavor caused by the formation of 3-methyl-2-butene-1-thiol when beer is exposed to ultraviolet light. They are effective only when unreduced (original) alpha acids and iso-alpha acids are not present. The perceived sensory bitterness of rho hop extracts is about 70 percent of traditional bittering compounds.

Beta-acids with hop oils. Brewers use these to control foaming in the kettle and increase the quantity of hop oils in wort.

Iso extracts. A standard solution of iso-alpha acids, usually sold in a concentration of 30 percent and added post-fermentation to adjust bitterness. These contain only traces of beta acids or essential oils.

Tetra extracts. Used to enhance foam and for light stability, tetrahydro-iso-alpha-acids are usually sold in a concentration of 10 percent. Like rho extracts, they prevent the development of light-struck flavors only in the absence of alpha acids and iso-alpha acids. Foam stand improves noticeably with the addition of 3 parts per million, which is below the threshold of most drinkers. The perceived bitterness, depending on beer type in which used, is about 1 to 1.7 times that of iso-alpha acids.

Hexa extracts. Hexahydro-iso-alpha-acids also provide foam enhancement and light stability, but their perceived bitterness is more similar to iso-alpha acids.

Hop aroma products. Each hop merchant offers a range of hop oil essences. They may be taken from cones or hop extracts. They may be general (such as "noble"), more specific ("citrusy"), or variety specific. In addition, through Botanix, the Barth-Haas group sells what it calls Pure Hop Aroma (PHA), both generic ("herbal, spicy, floral") and varietal, the latter in versions intended to emulate late hops or specifically dry hops

(Topnotes). Botanix developed Topnotes to enhance the dry-hop aroma of bottled beers, which were perceived as having less "hoppy" aroma than corresponding draft versions.

Botanix sales director Chris Daws said the Barth-Haas Hops Academy introduced many English craft brewers to Topnotes. Some use it to add aroma or flavor, while others add it to mask unwanted attributes in low alcohol beers. "You smell a Topnote, they don't smell like hops. Each Topnote is a concentrated form of a natural product. You have to put them in beer," he said.

That's the first thing he teaches potential customers. "They'll say, 'We want to get this in our beer.' I tell them we need to see what their beer is like before we can make a recommendation," he said. "All beers are individuals."

All hops are as well.

From Admiral to Zeus

The pages that follow provide only an introduction to 105 varieties of hops. As brewers emphasized in talking about selecting their hops each year, and will again in providing recipes, it is necessary to brew with a variety or a combination of varieties to get to know it/them.

That's why there are no suggested substitutions. Swapping one hop in for another is seldom straightforward. Lublin from Poland and Spalt Spalter are great hops and genetically nearly, if not totally, identical. Replace the Spalter hops in Stadtbrauerei Spalt *Pils* brewed in Spalt with Lublin, and the aroma and flavor may be excellent but likely a bit different. There is no one-on-one substitute for Amarillo, Centennial, or East Kent Golding. It would be silly to suggest otherwise. Brewers are better off determining the particular characteristics they want and identifying varieties with those.

Classifying hops. Sierra Nevada Brewing packs Magnum, generally considered a "bittering" hop, into the torpedoes used to dry hop *Torpedo Extra IPA*. The recipe for *Kellerpils* that Ron Barchet of Victory Brewing provides in Chapter 10 is bittered with Saaz and Hallertau Mittelfrüh, "aroma" hops. And, of course, there is the matter of "flavor hops" or hops with "special" aroma. Thus, it should be obvious that the letters beside the name of each variety are offered only as a broad (but friendly) guide.

- **B** *is for bittering* hops, used primarily because they have higher levels of alpha acids and thus are more efficient for bittering. Brewers may still be interested in their aroma attributes, particularly those looking for hops with higher percentages of essential oils to use in dry hopping.
- **N** *is for New World*, including American hops and those from the Southern Hemisphere, although new varieties from just about anywhere might qualify. American "C" hops, often beginning with that letter and usually floral and citrusy, belong here.
- **O** *is for organic*; that is, this variety is available as a certified organic hop.
- **P** *is for proprietary* or *patented* and includes trademarked varieties, most often hops developed by private breeding programs. Those interested in growing the varieties would need to acquire a license, as opposed to acquiring rhizomes for some hops in the public domain at nominal prices. Outside the United States, plant variety rights may apply, including restrictions.
- **S** *is for "special."* Perhaps somebody will coin a better word or at least provide a specific definition. "Flavor" is a contender. These are the varieties currently in fashion, new and often different, with bold aromas and flavors. They are a subset of what many call "dual purpose," hops with enough alpha acids to be used for bittering but also pleasant aroma qualities. There is no firm line between "special" and "almost special." For instance, demand for Centennial and Chinook has increased almost as quickly as that for Citra and Simcoe, but the former two are classified as New World, the latter as Special. Your mileage may vary.
- **T** *is for traditional*, primarily identifying landrace hops and those bred to provide similar aromas.

The numbers. Growing interest in oil content may change what information hop merchants routinely provide, although scientists must first establish with greater certainty the importance of compounds such as linalool and geraniol. Right now, they include data about the alpha and beta acids, cohumulone, total oil, and what percentage of the total oil each of the four main hydrocarbons constitute. Some publish a range, others a specific number, but brewers should be aware that large changes may occur from year to year, plus the variations that result from where hops are grown and when they are harvested.

New Zealand provides a floral-estery and citrus-piney measure of each of its varieties, listed with the descriptions that follow. The floral-estery component includes linalool, geranyl acetate, geranyl isobutyrate, and geraniol. The citrus-piney component includes limonene, d-cadinene, g-cadinene, muurolene, and selinine.

For the sake of comparison, here are a few non-New Zealand hops (measured in 1999), with the floral-estery component first: Hallertau Mittelfrüh 1.2/5.2, Tettnang Tettnanger 1.5/5.8, Nugget 1.5/5.8, and Magnum 1.1/3.4.

Storage. Hop processors use UV spectrophotometry to measure the Hop Storage Index (HSI) and include that on the labels of packages sent to commercial breweries. That allows brewers to calculate how quickly alpha acids will deteriorate and adjust their recipes as hops age. Catalogs may also, but do not always, provide an estimate of how well a particular variety stores. If they do, that's included here. Hops remain in much better condition when stored cold but can also be ruined before they ever get to the brewery (p. xx).

Why A to Zeus rather than A to ZYTHOS? ZYTHOS is not an actual hop. Hop supplier Hopunion first created a hop pellet blend it calls Falconer's Flight in 2010 and since has followed it up with Falconer's Flight 7C's and ZYTHOS. They include "special" varieties in short supply and in some cases even experimental hops. "I thought, 'How cool would it be to make an IPA in a hop,' " said Jesse Umbarger, a sales manager at Hopunion who helped set the blends. "We tried (for ZYTHOS) to hit an oil profile of Amarillo, Simcoe, Citra, and Centennial." Hopunion contracts for the blends just like any variety, with contracts that go five years forward. Umbarger said beginning with the 2012 harvest the company would provide the same information about oil content as it does for individual hops.

Admiral BOT

Bred in the United Kingdom, used primarily as an alternative to Target for bittering traditional English ales. Has fruity, citrus aroma that is more apparent when added later in the boil. An organic version is grown in Belgium. **Storage:** Good.

13-16%	alpha	1-1.7%	total oil
4.8-6.1%	beta	37-45%	cohumulone
39-48%	myrcene	1.8-2.2%	farnesene
23-36%	humulene	6.8-7.2%	caryophyllene

Agnus BT

Several large Czech breweries like Agnus as a bittering hop because its alpha/beta ratio is relatively high, which they believe makes their beer more stable. Herbal, grassy. High geraniol level, oils indicate dry hopping potential.

9-12%	alpha	2-3%	total oil
4-6.5%	beta	29-38%	cohumulone
40-55%	myrcene	<1%	farnesene
15-22%	humulene	6.8-7.2%	caryophyllene

Ahtanum NOP

Hits all the marks for American "C" hops, particularly its floral, spicy, herbal, and moderately piney aroma. Most often used for late hopping and dry hopping. Well established as an organic hop. **Storage:** Fair to good.

5.2-6.3%	alpha	0.8-1.2%	total oil
5-6.5%	beta	30-35%	cohumulone
50-55%	myrcene	<1%	farnesene
16-20%	humulene	9-12%	caryophyllene

Amarillo   NSP

Darren Gamache has suggested calling it a "landrace" hop because it was found growing wild (p. 81). Intensely fruity (citrus, melon, and stone fruits), well suited for American "hop bombs." Acreage increased significantly in 2012. **Storage:** Fair.

8-11%	alpha	1.5-1.9%	total oil
6-7%	beta	21-24%	cohumulone
68-70%	myrcene	2-4%	farnesene
9-11%	humulene	2-4%	caryophyllene

Apollo   BP

A patented hop from the S.S. Steiner breeding program. Used for bittering, although its pleasant aroma and oil profile make it a candidate for dry hopping along with a variety with bolder character. **Storage:** Excellent.

15-19%	alpha	1.5-2.5%	total oil
5.5-8%	beta	24-28%	cohumulone
30-50%	myrcene	<1%	farnesene
20-35%	humulene	14-20%	caryophyllene

Aramis   T

Recently released cross between classic, but low alpha, Strisselspalt variety unique to the French Alsace region and Whitbread Golding (both landrace hops). Spicy and herbal, perfumy floral notes, understated citrus. **Storage:** Fair.

7.9-8.3%	alpha	1.2-1.6%	total oil
3.8-4.5%	beta	21.5-21.7%	cohumulone
40%	myrcene	<1%	farnesene
21%	humulene	8%	caryophyllene

Aurora   T

From Slovenia. Previously known as Super Styrian, with many of the same traits as Styrian Golding, although it is a child of Northern Brewer. Higher percentage of alpha acids than Golding, more floral, spicier. **Storage:** Very good.

7-9%	alpha	0.9-1.4%	total oil
3-5%	beta	23-28%	cohumulone
35-53%	myrcene	6-9%	farnesene
20-27%	humulene	4.8%	caryophyllene

Bramling Cross   NT

Bramling, one of many Golding hops first mentioned in the mid-nineteenth century, was crossed in the U.K. with a brother of Brewer's Gold. Has black currant aroma, the "American tang" U.K. brewers described in 1930s. **Storage:** Poor.

6-7.8%	alpha	0.8-1.2%	total oil
2.2-2.8%	beta	26-31%	cohumulone
35-40%	myrcene	<1%	farnesene
28-33%	humulene	14-18%	caryophyllene

Bravo   BOP

A "super alpha" prized mostly for efficient bittering. Acreage increased steadily between 2007 and 2011. **Storage:** Good.

14-17%	alpha	1.6-2.4%	total oil
3-4%	beta	29-34%	cohumulone
25-50%	myrcene	<1%	farnesene
18-20%	humulene	10-12%	caryophyllene

Brewer's Gold   NT

Numbers are for United States, with alpha and oils lower in U.K. and lower still in Germany. Easier to find than Bullion but becoming an oddity. Aroma somewhat milder in Germany but in all cases still rich in black currant. **Storage:** Poor.

8-10%	alpha	2.2-2.4%	total oil
3.5-4.5%	beta	40-48%	cohumulone
37-40%	myrcene	<1%	farnesene
29-31%	humulene	7-7.5%	caryophyllene

Calypso   NP

A new, dual-purpose hop from the Hopsteiner breeding program that may establish itself as "special." Not as tropical as name might imply but rich in both stone fruits (pears and peaches, notes of cherry) and citrus. **Storage:** Good.

12-14%	alpha	1.6-2.5%	total oil
5-6%	beta	40-42%	cohumulone
30-45%	myrcene	<1%	farnesene
20-35%	humulene	9-15%	caryophyllene

Cascade (NZ)   NO

Cascade is different everywhere it is grown (in an increasing number of locations), and New Zealand illustrates that difference well. Citrus-piney fraction 5.9% Floral-estery fraction 2.8% **Storage:** Good.

6-8%	alpha	1.1%	total oil
5-5.5%	beta	37%	cohumulone
53-60%	myrcene	6%	farnesene
14.5%	humulene	5.4%	caryophyllene

Cascade (US)   NO

May not be classified as "special" but its bold, floral, citrusy aroma and flavor began to change the definition of "hoppy." Most widely grown American "aroma" hop. **Storage:** Very poor.

4.5-7%	alpha	0.7-1.4%	total oil
4.8-7%	beta	33-40%	cohumulone
45-60%	myrcene	3.7%	farnesene
8-13%	humulene	3.6%	caryophyllene

Celeia   T

Slovenian daughter of Styrian Golding. Shares some of the same aroma/flavor but bolder, citrusy (including grapefruit). Bred for lagers but works well for late hopping ales with fruity fermentation character. **Storage:** Good.

4.5-6%	alpha	0.6-1%	total oil
2.5-3.5%	beta	27-31%	cohumulone
27-33%	myrcene	3-7%	farnesene
20-35%	humulene	8-9%	caryophyllene

Centennial   NO

Has been called "Super Cascade," and recent demand has skyrocketed right along with increased sales of IPAs. Uniquely floral, perhaps because of the *cis*-rose compound found in few other varieties. **Storage:** Fair.

9.5-11.5%	alpha	1.5-2.3%	total oil
3.5-4.5%	beta	29-30%	cohumulone
45-55%	myrcene	<1%	farnesene
10-18%	humulene	5-8%	caryophyllene

Challenger OT

Bred and grown mostly in the U.K., its character synonymous with "English hops," maybe because in the 1970s (then) brewing giant Bass embraced it. Fruity and spicy. **Storage:** Very good to excellent.

6.5-8.5%	alpha	1-1.7%	total oil
4-4.5%	beta	20-25%	cohumulone
30-42%	myrcene	1-3%	farnesene
25-32%	humulene	8-10%	caryophyllene

Chelan BP

A daughter of Galena, bred at John I. Haas before merger with Select Botanicals Group. Interesting citrus aroma but used almost exclusively for bittering. **Storage:** Excellent.

12-14.5%	alpha	1.5-1.9%	total oil
8.5-9.8%	beta	33-35%	cohumulone
45-55%	myrcene	<1%	farnesene
12-15%	humulene	9-12%	caryophyllene

Chinook BNO

The piney, resinous aroma it delivers when used in dry hopping has become a hallmark of hopcentric American beers. Bred for bittering and still used for that, but now known for complex, fruity-piney contributions. **Storage:** Good.

12-14%	alpha	1.7-2.7%	total oil
3-4%	beta	29-35%	cohumulone
35-40%	myrcene	<1%	farnesene
18-23%	humulene	9-11%	caryophyllene

Citra   **NOPS**

Poster child for "flavor" or "special" hops, and in demand well beyond the United States. Rich in passion fruit, lychee, peach, gooseberries, and a laundry list of other unusual (for hops) flavors. **Storage:** Fair.

11-13%	alpha	2.2-2.8%	total oil
3.5-4.5%	beta	22-24%	cohumulone
60-65%	myrcene	0%	farnesene
11-13%	humulene	6-8%	caryophyllene

Cluster   **NO**

As close as America comes to a landrace hop. Once accounted for 80 percent of hops grown in the Northwest. The organic version is a variety that was known as California Ivanhoe, a reminder of how many variations there likely were. **Storage:** Excellent.

5.5-8.5%	alpha	0.4-0.8%	total oil
4.5-5.5%	beta	37-43%	cohumulone
45-55%	myrcene	<1%	farnesene
15-18%	humulene	6-7%	caryophyllene

Columbus   **BNP**

One of 3 (along with Tomahawk and Zeus) varieties so genetically similar they are lumped together as "CTZ." Often used for bittering, but aromas differ. Columbus often brightly fruity and spicy. **Storage:** Very poor.

14-16.5%	alpha	2-3%	total oil
4-5%	beta	28-32%	cohumulone
40-50%	myrcene	<1%	farnesene
12-18%	humulene	9-11%	caryophyllene

Crystal   NT

As a progeny of Hallertau Mittelfrüh, Cascade, and Brewer's Gold, not surprising it takes on different character depending on how it is used. Can be mild, spicy, and floral, or in a beer like Rogue *Brutal IPA*, quite pungent. **Storage:** Very poor.

3.5-5.5%	alpha	1-1.5%	total oil
4.5-6.5%	beta	20-26%	cohumulone
40-65%	myrcene	<1%	farnesene
18-24%	humulene	4-8%	caryophyllene

Delta   NP

A backcross between Fuggle and Cascade, recently released by the Hopsteiner breeding program. Woody and herbal with a citrus kick. Not as assertive as dual purpose hops but adds distinctive notes in a blend. **Storage:** Excellent.

5-5.7%	alpha	0.5-1.1%	total oil
5.5-7%	beta	22-24%	cohumulone
25-40%	myrcene	<1%	farnesene
30-40%	humulene	9-15%	caryophyllene

El Dorado   NSP

Bred on a single Yakima Valley farm (p. 74), available on a larger scale for the first time after 2012 harvest. Fits the dual purpose/special mold perfectly. Intense aromas: stone fruits (pear, cherries), candy (Lifesavers). **Storage:** Good.

14-16%	alpha	2.5-2.8%	total oil
7-8%	beta	28-33%	cohumulone
55-60%	myrcene	<1%	farnesene
10-15%	humulene	6-8%	caryophyllene

Ella   NPS

Originally called Stella (which a large brewery took exception to). New name after 2013 harvest in Australia. Despite high alpha sold as an aroma variety. Bold but not in a tropical way, sweet and floral, with lemon, apricot, and melon.

14-16%	alpha	2.9%	total oil
4-4.5%	beta	36%	cohumulone
33%	myrcene	13%	farnesene
1.2%	humulene	15%	caryophyllene

First Gold   OT

Important because it is a true dwarf hop. Popular because it retained much of the fruity, floral character of its mother, Whitbread Golding. Very British, works well for all kettle additions as well as dry hopping. **Storage:** Very good.

5.6-9.3%	alpha	0.7-1.5%	total oil
2.3-4.1%	beta	32-34%	cohumulone
24-27%	myrcene	2-4%	farnesene
20-24%	humulene	6-7%	caryophyllene

Fuggle   OT

Pretty well defines English hop character (fruity, spicy, woody) by itself or along with Golding. Future cloudy because of agronomic weaknesses, but important ancestor to many modern hops. **Storage:** Fair.

3-5.6%	alpha	0.7-1.4%	total oil
2-3%	beta	25-30%	cohumulone
24-28%	myrcene	5-7%	farnesene
33-38%	humulene	9-13%	caryophyllene

Galaxy   NPS

Australian variety that helped inspire the term "flavor hop." Hign in alpha but used mostly for late/dry hopping. Rich in passion fruit, citrus, apricot, melon, black currant. Can be intense, even pungent.

13.5-15%	alpha	2.4-2.7%	total oil
5.8-6%	beta	35%	cohumulone
33-42%	myrcene	3-4%	farnesene
1-2%	humulene	9-12%	caryophyllene

Galena   B

A popular high alpha hop since it was released in 1978, but acreage dropped after 2007, some of it replaced by Super Galena. Occasionally used for its floral/citrus aroma at large breweries focused on pale lagers. **Storage:** Very good to excellent.

11-13.5%	alpha	0.9-1.3%	total oil
7.2-8.7%	beta	36-40%	cohumulone
55-60%	myrcene	<1%	farnesene
10-13%	humulene	4.5-6.5%	caryophyllene

Glacier   T

A product of the USDA breeding program, notable for extremely low cohumulone. Pleasant, classically mild aroma, floral, including citrus and stone fruits, most notably peach. **Storage:** Good.

5-6%	alpha	0.7-1.6%	total oil
7.6%	beta	11-13%	cohumulone
33-62%	myrcene	<1%	farnesene
24-36%	humulene	6.5-10%	caryophyllene

Green Bullet   BNO

Although known for its then-high alpha when released in 1972, now valued for its aroma. Floral and fruity (including hint of grape). Floral-estery fraction 2.3% Citrus-piney fraction 7.9% **Storage:** Good.

11-14%	alpha	1.1%	total oil
6.5-7%	beta	41-43%	cohumulone
38%	myrcene	<1%	farnesene
28%	humulene	9%	caryophyllene

Hallertau Merkur   B

A bittering hop used primarily in pale lagers. Earthy, floral, spicy. **Storage:** Good.

10-14%	alpha	1.4-1.9%	total oil
3.5-7%	beta	17-22%	cohumulone
25-35%	myrcene	<1%	farnesene
35-50%	humulene	9-15%	caryophyllene

Hallertau Mittelfrüh   OT

Classic landrace hop, complex but subtle, well suited for lagers. The USDA program bred numerous hybrids attempting to replicate its herbal, spicy, mildly woody aromas and flavors. Each is interesting, but none is Mittelfrüh. **Storage:** Poor.

3-5.5%	alpha	0.7-1.3%	total oil
3-5%	beta	18-28%	cohumulone
20-28%	myrcene	<1%	farnesene
45-55%	humulene	10-15%	caryophyllene

Hallertau Taurus

B

Released in Germany about the same time as Magnum but has not proved to be as popular. Works best in moderately hopped pale lagers, accenting its mild, traditional aroma. **Storage:** Good.

12-17%	alpha	0.9-1.4%	total oil
4-6%	beta	20-25%	cohumulone
30-50%	myrcene	<1%	farnesene
23-33%	humulene	6-11%	caryophyllene

Hallertau Tradition

OT

Bred in Germany to replicate the character of Mittelfrüh but with better yield and resistance to disease. Like its ancestor floral, herbal, and subtle. **Storage:** Good.

4-7%	alpha	0.5-1%	total oil
3-6%	beta	24-30%	cohumulone
17-32%	myrcene	<1%	farnesene
35-50%	humulene	10-15%	caryophyllene

Harmonie

T

With "typical" Czech aroma, Harmonie's finely balanced bitterness "persists" slightly longer than Saaz. Bohemie, released only in 2010, is a similar Czech hop with traditional aroma and 5-8% alpha acids.

5-8%	alpha	1-2%	total oil
5-8%	beta	17-21%	cohumulone
30-40%	myrcene	<1%	farnesene
10-20%	humulene	6-11%	caryophyllene

Helga   PT

Previously known as Southern Hallertau, an offspring of Hallertau Mittelfrüh bred in Australia. Same spicy, herbal qualities as Mittelfrüh.

5-6%	alpha	0.6-0.7%	total oil
3.8-5.4%	beta	22-26%	cohumulone
2-12%	myrcene	<1%	farnesene
35-47%	humulene	10-14%	caryophyllene

Herkules   B

German bred, aptly named, Herkules looks like an alpha tree in the field, thick with cones from the ground to the wire. High alpha, high yields. Smoothly bitter, a reminder that assessing cohumulone's role is complicated. **Storage:** Very good.

12-17%	alpha	1.6-2.4%	total oil
4-5.5%	beta	32-38%	cohumulone
30-50%	myrcene	<1%	farnesene
30-45%	humulene	7-12%	caryophyllene

Hersbrucker   OT

Or Hersbrucker Spät. Hersbruck, north of the Hallertau, was once one of Germany's main hop growing regions. Classically spicy, herbal, with hints of citrus, stone fruit. Would make a perfect hop perfume. **Storage:** Poor to fair.

1.5-4%	alpha	0.5-1%	total oil
2.5-6%	beta	17-25%	cohumulone
15-30%	myrcene	<1%	farnesene
20-30%	humulene	8-13%	caryophyllene

Horizon   BN

A half-sister of Nugget that some brewers like for bittering because of its low cohumulone. Dual purpose because of its floral character and spicy undertones. **Storage:** Fair to good.

11-13%	alpha	1.5-2%	total oil
6.5-8.5%	beta	16-19%	cohumulone
55-65%	myrcene	2.5-3.5%	farnesene
11-13%	humulene	2.5-3.5%	caryophyllene

Kazbek   T

The Czechs bred Kazbek to be tolerant of drought and hot weather, including wild hops of Russian origin in the crosses. Its strong, spicy aroma sets it apart from other Czech hops.

5-8%	alpha	0.9-1.8%	total oil
4-6%	beta	35-40%	cohumulone
40-50%	myrcene	<1%	farnesene
20-35%	humulene	10-15%	caryophyllene

Kent Golding   T

Designating The Original Golding is no easier than sorting out the origins of the Variety (with a capital *V*). Even East Kent differs from Kent and certainly from American grown. Most important, it tastes of English beer. **Storage:** Very good.

4-6.5%	alpha	0.4-0.8%	total oil
1.9-2.8%	beta	28-32%	cohumulone
20-26%	myrcene	<1%	farnesene
38-44%	humulene	12-16%	caryophyllene

Liberty   OT

One of four triploid Hallertau Mittelfrüh varieties released by the USDA program, and considered "closest" to the original. **Storage:** Very poor.

3-5%	alpha	0.6-1.2%	total oil
3-4%	beta	24-30%	cohumulone
20-40%	myrcene	<1%	farnesene
35-40%	humulene	9-12%	caryophyllene

Lublin   T

Grown in Poland and very similar to other Saazer-type landrace varieties. Also known as Lubliner or Lubelski. **Storage:** Very poor.

3-4.5%	alpha	0.5-1.1%	total oil
3-4%	beta	25-28%	cohumulone
22-29%	myrcene	10-14%	farnesene
30-40%	humulene	6-11%	caryophyllene

Magnum   BO

Soon to be replaced as Germany's primary high alpha hop (Herkules already yields more pounds, although Magnum is planted on more acres). Not surprisingly, German-grown version smells more "noble" than American. **Storage:** Very good.

11-16%	alpha	1.6-2.6%	total oil
5-7%	beta	21-29%	cohumulone
30-45%	myrcene	<1%	farnesene
30-45%	humulene	8-12%	caryophyllene

Mandarina Bavaria   NST

One of four new varieties released to German growers in 2012, the first products from the new German breeding imperative. Shares characteristics of its mother, Cascade, but fruitier and more herbal.

7-10%	alpha	2.1%	total oil
5-7%	beta	33%	cohumulone
71%	myrcene	1%	farnesene
5%	humulene	1.7%	caryophyllene

Marynka   BT

Primary bittering hop in Poland but with enough aroma to be considered dual purpose. Very aromatic, hinting of freshly picked flowers, some of them roses.

6-12%	alpha	1.8-2.2%	total oil
10-13%	beta	26-33%	cohumulone
28-31%	myrcene	1.8-2.2%	farnesene
26-33%	humulene	11-12%	caryophyllene

Meridian   N

Indie Hops planned to revive Columbia, the unplanted sister of Willamette. However, the hill at Goschie Farms thought to contain Columbia yielded this previously unknown hop. Interesting aroma includes lemon pie, fruit punch.

6.5%	alpha	1.1%	total oil
9.5%	beta	45%	cohumulone
30%	myrcene	<1%	farnesene
8%	humulene	3.8%	caryophyllene

Millennium   B

A daughter of Nugget, valued primarily for its alpha, but very pleasant, spicy aroma. The sommeliers who contributed to the Barth *Hop Aroma Compendium* noticed "creme-caramel flavors of yogurt and toffee." **Storage:** Excellent.

14-16.5%	alpha	1.8-2.2%	total oil
4.3-5.3%	beta	28-32%	cohumulone
30-40%	myrcene	<1%	farnesene
23-27%	humulene	9-12%	caryophyllene

Mosaic   NS

Available in quantity for the first time after 2012 harvest. Still known to many as HBC 369. A daughter of Simcoe crossed with a disease-resistant, Nugget-derived male. Rich in mango, lemon, citrus, pine, and, notably, blueberry.

11-13.5%	alpha	1.5%	total oil
3.2-3.9%	beta	2.4-2.6%	cohumulone
54%	myrcene	<1%	farnesene
13%	humulene	6.4%	caryophyllene

Motueka   NOS

One-third Saaz crossed with New Zealand stock. First called Belgian Saaz. Citrus, notably lemon and lime, and tropical fruit. Floral-estery fraction 4% Citrus-piney fraction 18.3% **Storage:** Good.

6.5-7.5%	alpha	0.8%	total oil
5-5.5%	beta	29%	cohumulone
48%	myrcene	12%	farnesene
3.6%	humulene	2%	caryophyllene

Mt. Hood   T

Another triploid daughter of Hallertau Mittelfrüh. Flourished in the 1990s but limited acreage today. **Storage: Poor.**

4-7%	alpha	1.2-1.7%	total oil
5-8%	beta	21-23%	cohumulone
30-40%	myrcene	<1%	farnesene
30-38%	humulene	13-16%	caryophyllene

Nelson Sauvin   NOS

Much in demand because of its Sauvignon Blanc character. Richly fruity, somewhat tropical, beyond the white wine notes. Floral-estery fraction 2.8% Citrus-piney fraction 7.8% **Storage:** Good.

12-13%	alpha	1-1.2%	total oil
6-8%	beta	22-26%	cohumulone
21-23%	myrcene	<1%	farnesene
35-37%	humulene	10-12%	caryophyllene

Newport   BN

A high alpha hop developed by the USDA. Overall aroma is mild, but used in dry hopping can be pungent, resiny. **Storage:** Fair.

13.5-17%	alpha	1.6-3.4%	total oil
7.2-9.1%	beta	36-38%	cohumulone
47-54%	myrcene	<1%	farnesene
9-14%	humulene	4.5-7%	caryophyllene

Northdown T

Notable in the 1970s for its relatively high alpha, since supplanted by other varieties. Bitterness leans toward harsh, but its sometimes high oils make it useful for dry hopping. Relatively neutral, but English, character. **Storage:** Fair to good.

7.5-9.5%	alpha	1.2-2.5%	total oil
5-5.5%	beta	24-30%	cohumulone
23-29%	myrcene	<1%	farnesene
40-45%	humulene	13-17%	caryophyllene

Northern Brewer NT

An offspring of Brewer's Gold, with high enough alpha that it was once a popular bittering hop in Germany. Milder aroma (half the level of myrcene and little "American tang") when grown in Germany. **Storage:** Very good to excellent.

6-10%	alpha	1-1.6%	total oil
3-5%	beta	27-32%	cohumulone
50-65%	myrcene	<1%	farnesene
35-50%	humulene	10-20%	caryophyllene

Nugget BO

Released by the USDA in 1983 to meet demand for higher alpha hops, and remains a staple for Oregon farmers. Pleasant herbal aroma. Grown in Germany (again with milder aroma) for 20 years but not much now. **Storage:** Very good to excellent.

11-14%	alpha	0.9-2.2%	total oil
3-5.8%	beta	22-30%	cohumulone
48-55%	myrcene	<1%	farnesene
16-19%	humulene	7-10%	caryophyllene

Opal   OT

Developed at German Hop Research Center Hüll, distinctive because of its spicy, woody aromatics. Moderately floral. **Storage:** Fair.

5-8%	alpha	0.8-1.3%	total oil
3.5-5.5%	beta	13-17%	cohumulone
20-45%	myrcene	<1%	farnesene
30-50%	humulene	8-15%	caryophyllene

Pacifica   NO

A cross between Hallertau Mittelfrüh and New Zealand breeding stock, formerly called Pacific Hallertau. Basically a bolder version of Mittelfrüh. Floral-estery fraction 1.6% Citrus-piney fraction 6.9% **Storage:** Good.

5-6%	alpha	1%	total oil
5.5-6%	beta	25%	cohumulone
10-14%	myrcene	<1%	farnesene
48-52%	humulene	17%	caryophyllene

Palisade   NOP

Bred at Yakima Chief Ranches primarily for aroma (floral, fruity, and can be tropical). Versatile, complements hops with "special" aroma well. **Storage:** Good.

5.5-9.5%	alpha	1.4-1.6%	total oil
6-8%	beta	24-29%	cohumulone
9-10%	myrcene	<1%	farnesene
19-22%	humulene	16-18%	caryophyllene

Perle OT

Bred in Germany with Northern Brewer as a parent. Minty, spicy aroma, and what Germans call "mild." Some years alpha high enough to use as bittering hop. Also grown in American Northwest. **Storage:** Very good to excellent.

4-9%	alpha	0.5-1.5%	total oil
2.5-4.5%	beta	29-35%	cohumulone
20-35%	myrcene	<1%	farnesene
35-55%	humulene	10-20%	caryophyllene

Pilgrim BO

Although it has the same father as U.K. dwarf hops First Gold and Herald, it is not a dwarf. Classified a bittering hop by some, dual purpose by others, has interesting lemon and grapefruit aroma. **Storage:** Very good.

9-13%	alpha	1.2-2.4%	total oil
4.3-5%	beta	36-38%	cohumulone
30%	myrcene	0.3%	farnesene
17%	humulene	7.3%	caryophyllene

Pioneer T

Bred in England for low trellises and considered dual purpose. Underlying citrus aroma similar to Golding varieties, complemented by hint of lemon. **Storage:** Good.

7-11%	alpha	1-1.8%	total oil
3.5-4%	beta	36%	cohumulone
31-36%	myrcene	<1%	farnesene
22-24%	humulene	7-8%	caryophyllene

Polaris   BNT

Another of the new varieties available to German hop growers beginning in 2012. Very high alpha and stunningly high oil content, its aroma variously described as "ice candy," eucalyptus, peppermint, and citrus.

19-23%	alpha	4.4%	total oil
5-7%	beta	27%	cohumulone
50%	myrcene	<0.1%	farnesene
22%	humulene	9%	caryophyllene

Premiant   T

Bred in the Czech Republic from selected Saaz progenies, resulting in a higher percentage of alpha acids. Aroma profile is clean, floral, and slightly citrus. In trials tasters found its bitterness especially soft and well rounded.

7-9%	alpha	1-2%	total oil
3.5-5.5%	beta	18-23%	cohumulone
35-45%	myrcene	1-3%	farnesene
25-35%	humulene	7-13%	caryophyllene

Pride of Ringwood   B

A noteworthy Australian bittering hop that established itself in the 1960s. Its alpha levels were surpassed long ago, but it has some of the aroma characteristics (berry flavors, citrus) now in vogue. **Storage:** Very poor.

7-11%	alpha	0.9-2%	total oil
4-6%	beta	32-39%	cohumulone
25-50%	myrcene	<1%	farnesene
2-8%	humulene	5-8%	caryophyllene

Progress   T

Bred in the 1960s in England as an aroma hop but has a character that makes it seem as if it could be a landrace variety. Well suited to English-style ales. **Storage:** Fair.

5-7%	alpha	0.6-1.2%	total oil
2-2.5%	beta	25-30%	cohumulone
30-35%	myrcene	<1%	farnesene
40-47%	humulene	12-15%	caryophyllene

Rakau   NO

Previously known as Alpharoma. Has been compared to Nelson Sauvin as a "flavor hop." Aroma/flavor includes tropical fruit, passion fruit, and peach. Floral-estery fraction 1.2% Citrus-piney fraction 5.7% **Storage:** Good.

10.8%	alpha	2.1%	total oil
4.6%	beta	25%	cohumulone
56%	myrcene	4.5%	farnesene
16.3%	humulene	5.2%	caryophyllene

Riwaka   NO

Saazer-type hop crossed with New Zealand breeding material, first called D Saaz. Piney, tropical aromas.In short supply. Floral-estery fraction 2.8% Citrus-piney fraction 5.9% **Storage:** Good.

4.5-6.5%	alpha	0.8%	total oil
4-5%	beta	29-36%	cohumulone
68%	myrcene	1%	farnesene
9%	humulene	4%	caryophyllene

Rubin   B

The Czechs classify this as a bittering hop, but with its Saaz background it is genetically similar to European aroma varieties. Bitterness is not as smooth as Saaz, and it lingers longer.

9-12%	alpha	1-2%	total oil
3.5-5%	beta	25-33%	cohumulone
35-45%	myrcene	<1%	farnesene
13-20%	humulene	7-10%	caryophyllene

Saaz   OT

Although farmers cut Saaz acreage in 2011, it still accounts for 83 percent of Czech hops planted. Grown elsewhere but original distinctive, pleasant, and delicate. First organic Saaz released after 2012 harvest. **Storage:** Poor.

3-6%	alpha	0.4-1%	total oil
4.5-8%	beta	23-26%	cohumulone
25-40%	myrcene	14-20%	farnesene
15-25%	humulene	10-12%	caryophyllene

Saaz Late   T

Bred at Czech Hop Research Institute and developed to replicate classic Saaz aroma but reduce yearly swings in levels of alpha acids.

3-7%	alpha	0.5-1%	total oil
3.8-6.8%	beta	20-24%	cohumulone
25-35%	myrcene	15-20%	farnesene
15-20%	humulene	6-9%	caryophyllene

Santium   T

Developed by the USDA in Oregon to mimic the character of Tettnang Tettnanger with high level of alpha acids. Herbal and spicy. **Storage:** Fair to good.

5.5-7%	alpha	1.3-1.7%	total oil
7.8.5%	beta	20-22%	cohumulone
30-45%	myrcene	13-16%	farnesene
20-25%	humulene	5-8%	caryophyllene

Saphir   OT

From the German hop breeding program, developed like Opal and Smaragd for "classic" aroma. Pleasantly spicy, with hints of New World berry/citrus character. Versatile, stands up to fruity/clovy ale yeasts. **Storage:** Good.

2-4.5%	alpha	0.8-1.4%	total oil
4-7%	beta	12-17%	cohumulone
25-40%	myrcene	<1%	farnesene
20-30%	humulene	9-14%	caryophyllene

Simcoe   NOSP

Its aroma has become another hallmark of dry-hopped American beers, pushed to the edge of pungent and "catty" and sometimes beyond. Intense, rich in multiple citrus fruits, black currant, berries, and pine. **Storage:** Good.

12-14%	alpha	2-2.5%	total oil
4-5%	beta	15-20%	cohumulone
60-65%	myrcene	<1%	farnesene
10-15%	humulene	5-8%	caryophyllene

Sládek   T

Like Premiant, bred from selected Saaz progenies and has percentage of alpha acids closer to the original. Floral and slightly spicy, rated highly for its flavor and overall balance.

4.5-6.5%	alpha	1-2%	total oil
4-6%	beta	25-30%	cohumulone
40-50%	myrcene	<1%	farnesene
20-30%	humulene	8-13%	caryophyllene

Smaragd   OT

Smaragd means "emerald" in German, and, like Opal and Saphir, it was bred at Hüll for traditional hop aroma and flavor. Spicy, herbal, and woody. **Storage:** Fair.

4-6%	alpha	0.4-0.8%	total oil
3.5-5.5%	beta	13-18%	cohumulone
20-40%	myrcene	<1%	farnesene
30-50%	humulene	9-14%	caryophyllene

Sorachi Ace   NS

Bred in Japan, with both Saaz and Brewer's Gold in its background, was not being grown anywhere until Darren Gamache claimed it out of the USDA archives. Striking lemon character, a bold hop for bold beers.

10-16%	alpha	2-2.8%	total oil
6-7%	beta	23%	cohumulone
35%	myrcene	6%	farnesene
21-27%	humulene	8-9%	caryophyllene

Southern Cross NO

Dual purpose. Popular in New Zealand lagers. Has attractive aromas that include citrus (lemon), spice, and pine. Floral-estery fraction 2.7% Citrus-piney fraction 6.9% **Storage:** Good.

11-14%	alpha	1.2%	total oil
5-6%	beta	25-28%	cohumulone
32%	myrcene	7.3%	farnesene
21%	humulene	6.7%	caryophyllene

Sovereign OT

Another English dwarf hop, a daughter of Whitbread Golding. Although its aroma can be intensely fruity, Sovereign also produces softer flavors, including stone fruits such as peach.

4.5-6.5%	alpha	0.8%	total oil
2.1-3.1%	beta	26-30%	cohumulone
not provided	myrcene	3.6%	farnesene
23%	humulene	8.3%	caryophyllene

Spalt Spalter T

Available only from the Spalt region, little about growing it is easy but its *fine* aroma is unique despite genetic similarities to Saaz, Tettnanger. Spicy, delicate, herbal, woody, floral. **Storage:** Poor.

2.5-5.5%	alpha	0.5-0.9%	total oil
3-5%	beta	22-29%	cohumulone
20-35%	myrcene	12-18%	farnesene
20-30%	humulene	8-13%	caryophyllene

Spalter Select   OT

Much more widely grown in Germany than the original Spalt, and almost as popular as Mittelfrüh. Developed at Hüll, spicy, floral, and woody. A solid replacement for Saazer-type hops. **Storage:** Fair.

3-6.5%	alpha	0.6-0.9%	total oil
2.5-5%	beta	21-27%	cohumulone
20-40%	myrcene	15-22%	farnesene
10-22%	humulene	4-10%	caryophyllene

Sterling   OT

The daughter of a Saaz clone and a father that had Cascade and several European cultivars in his background. The result is a hop with Saaz character, including a spicy and citrus aroma, which has much more alpha acid content. **Storage:** Good.

6-9%	alpha	1.3-1.9%	total oil
4-6%	beta	22-28%	cohumulone
44-48%	myrcene	11-17%	farnesene
19-23%	humulene	5-7%	caryophyllene

Strisselspalt   T

Landrace hop grown in French Alsace. Acreage almost disappeared after Anheuser-Busch pulled contracts. Elegant aroma, floral, spicy, and lemon zest. Once a staple in Michelob pale lager but matches Belgian yeasts well. **Storage:** Fair.

1.8-2.5%	alpha	0.6-0.8%	total oil
3-6%	beta	20-25%	cohumulone
35-52%	myrcene	<1%	farnesene
12-21%	humulene	6%-10%	caryophyllene

Styrian Golding   T

Looking for a replacement hop because their fields were devastated by disease, Slovenian farmers brought home what they thought was a Golding in the 1930s and called it Savinja Golding. It was a Fuggle. Now delicately different in Slovenia. **Storage:** Very good.

4.5-6%	alpha	0.5-1%	total oil
2-3.5%	beta	25-30%	cohumulone
27-33%	myrcene	3-5%	farnesene
20-35%	humulene	7-10%	caryophyllene

Summer   PT

Summer and Sylva are Australian sisters that were selected from Saaz crosses, with chemical characteristics similar to their ancestor. Summer is lighter and fruitier, though still spicy and floral, with tealike components.

4-7%	alpha	0.9-1.3%	total oil
4.8-6.1%	beta	22-25%	cohumulone
5-13%	myrcene	<1%	farnesene
42-46%	humulene	14-15%	caryophyllene

Summit   BNOP

A low-trellis hop (not a true dwarf) developed in the Northwest. Ancestors include Zeus and Nugget. Has strong citrus and grapefruit aromas and flavors, making it suitable as a dual purpose hop, but can drift toward onion and garlic. **Storage:** Excellent.

13-15.5%	alpha	1.5-2.5%	total oil
4-6%	beta	26-33%	cohumulone
30-50%	myrcene	<1%	farnesene
15-25%	humulene	10-15%	caryophyllene

Super Galena   BP

A granddaughter of Galena. Noteworthy for both high alpha and yield. Can produce more alpha per acre than even Herkules.

13-16%	alpha	1.5-2.5%	total oil
8-10%	beta	35-40%	cohumulone
45-60%	myrcene	<1%	farnesene
19-24%	humulene	6-14%	caryophyllene

Super Pride   BP

Australian offspring of Pride of Ringwood, similar but with higher levels of alpha acids. **Storage:** Good.

14-15%	alpha	1.7-1.9%	total oil
7-8%	beta	30-34%	cohumulone
30-45%	myrcene	<1%	farnesene
1-2%	humulene	7-9%	caryophyllene

Sylva   PT

A sister to Summer, Sylva displays what's considered the traditional hoppy (with a cedarwood note) and spicy character of Saaz. Mostly an option for Southern Hemisphere brewers but an interesting variation on Saaz.

4-7%	alpha	0.5-1.1%	total oil
3-5%	beta	23-28%	cohumulone
17-23%	myrcene	23-25%	farnesene
19-26%	humulene	6-9%	caryophyllene

Target   BT

Target quickly became the most grown hop in England shortly after it was released in 1972, reflecting what would be a worldwide trend toward high alpha hops. Still used by brewers focused on English character. Interesting as a dry hop. **Storage:** Poor.

9.5-12.5%	alpha	1.2-1.4%	total oil
4.3-5.7%	beta	35-40%	cohumulone
45-55%	myrcene	<1%	farnesene
17-22%	Humulene	8-10%	caryophyllene

Tettnang Tettnanger   OT

A member of the Saazer family, with similar but different aromas, unique to the Tettnang region. Descriptors from the Barth *Hop Aroma Compendium*: "Floral, bergamot, lily of the valley, cognac, chocolate." **Storage:** Poor.

2.5-5.5%	alpha	0.5-0.9%	total oil
3-5%	beta	22-28%	cohumulone
20-35%	myrcene	16-24%	farnesene
22-32%	humulene	6-11%	caryophyllene

Tettnanger (US)   T

There are various theories about what happened, but this hop has more in common with Fuggle than it has with Tettnang Tettnanger. Reason to question the genetics of hops bred with U.S. Tettnanger. Woody, spicy. **Storage:** Good.

4-5%	alpha	0.4-0.8%	total oil
3.5-4.5%	beta	20-25%	cohumulone
25-40%	myrcene	10-15%	farnesene
18-25%	humulene	6-8%	caryophyllene

Tillicum   BP

A daughter of Galena with pleasant aroma but best used only for bittering. Farmers grow it in order to stagger harvest dates with other high alpha varieties. **Storage:** Excellent.

12-14.5%	alpha	1.5-1.9%	total oil
9.3-10.5%	beta	31-38%	cohumulone
45-55%	myrcene	<1%	farnesene
13-16%	humulene	7-8%	caryophyllene

Tomahawk   BP

Another of the CTZ hops, and like the others has interesting features when considered as an individual. Researchers in Belgium recently found compounds in Tomahawk similar to those in Nelson Sauvin. **Storage:** Very poor.

14.5-17%	alpha	2.5-3.5%	total oil
4.5-5.5%	beta	28-35%	cohumulone
50-60%	myrcene	<1%	farnesene
9-15%	humulene	4-10%	caryophyllene

Topaz   BNP

A high alpha variety out of Australia, bred to go directly to extraction. Has since drawn interest as a dual purpose hop because of strong, fruity flavors, both berries and passion fruit. **Storage:** Excellent.

15-18%	alpha	0.8-1.7%	total oil
6-7%	beta	47-50%	cohumulone
25-43%	myrcene	<1%	farnesene
11-13%	humulene	10-11%	caryophyllene

Triskel   T

Newest release bred for Alsace hop growing region, a cross between French Strisselspalt and the English variety Yeoman. Mild enough to suit a pale lager but with an oil profile (floral, citrus) that fits American-style ales. **Storage:** Fair.

8-9%	alpha	1.5-2%	total oil
4-4.7%	beta	20-23%	cohumulone
60%	myrcene	<1%	farnesene
13.5%	humulene	5.4%	caryophyllene

Tsingtao Flower   N

Accounts for about 65 percent of acreage in China, the Chinese version of American Cluster. Floral and spicy. **Storage:** Good.

6-8%	alpha	0.4-0.8%	total oil
3-4.2%	beta	35%	cohumulone
45-55%	myrcene	<1%	farnesene
15-18%	humulene	6-7%	caryophyllene

Ultra   T

Bred in Oregon from Hallertau Mittelfrüh and a Saazer-type male. Another choice for those looking for the mild, pleasant character of European landrace hops. Loved by some brewers but never grown on many acres. **Storage:** Good to very good.

2-3.5%	alpha	0.5-1%	total oil
3-4.5%	beta	23-38%	cohumulone
15-25%	myrcene	<1%	farnesene
35-50%	humulene	10-15%	caryophyllene

Vanguard   T

Another offspring of Mittelfrüh developed by the USDA. Herbal and spicy, much like her mother. **Storage:** Very good to excellent.

5.5-6%	alpha	0.9-1.2%	total oil
6-7%	beta	14-16%	cohumulone
20-25%	myrcene	<1%	farnesene
45-50%	humulene	12-14%	caryophyllene

Wakatu   NOST

Its background includes two-thirds Hallertau Mittelfrüh, which is reflected in the aroma. Alpha is high enough to consider it dual purpose. Particularly bold. Floral-estery fraction 3.2% Citrus-piney fraction 9.5% **Storage:** Good.

6.5-8.5%	alpha	0.9-1.1%	total oil
8-9%	beta	28-30%	cohumulone
35-36%	myrcene	6-7%	farnesene
16-17%	humulene	7-9%	caryophyllene

Warrior   BP

Emerged from the same crosses that produced Simcoe, and likewise open pollinated. Used primarily for bittering but has interesting aroma/flavor attributes: floral, spicy, woody, and sweet citrus. **Storage:** Very good.

14-16.5%	alpha	1.3-1.7%	total oil
4.3-5.3%	beta	22-26%	cohumulone
40-50%	myrcene	<1%	farnesene
15-19%	humulene	9-11%	caryophyllene

Whitbread Golding   T

Not a true Golding but similar in character, with pronounced sweet fruitiness. Was bred and grown on the Whitbread Brewery Hop Farm in Kent, now a hop museum and amusement park. **Storage:** Fair.

5.4-7.7%	alpha	0.9-1.4%	total oil
2-2.5%	beta	25-36%	cohumulone
43%	myrcene	1-3%	farnesene
29-44%	humulene	12-14%	caryophyllene

Willamette   T

The most grown American aroma hop until Anheuser-Busch InBev cut back commitment in 2008. An alternative to Fuggle released in 1976, with a mild, spicy profile. Versatile, its flavor works well with many styles. **Storage:** Fair.

4-6%	alpha	1-1.5%	total oil
3-4.5%	beta	30-35%	cohumulone
30-40%	myrcene	5-6%	farnesene
20-27%	humulene	7-8%	caryophyllene

Zeus   BP

The third member of the CTZ clan. Like the others very aromatic, sometimes to the point of being pungent. Citrus notes are most apparent, but also has spicy and herbal character. **Storage:** Poor.

12-16.5%	alpha	1-2%	total oil
4-6%	beta	27-35%	cohumulone
25-65%	myrcene	<1%	farnesene
10-25%	humulene	5-15%	caryophyllene

7

Hops in the Brewhouse

Perception matters: You can have your bitterness and smell the aroma, too

The guest rooms at *Brauerei und Gasthof zur Krone* in Tettnang in southwest Germany are thoroughly modern, with hardwood floors, whitewashed walls, and the sleek amenities tourists spending their holiday in the Lake Constance (Bodensee) region expect. The building, on the other hand, has been around since before the last Montfort, Count Anton IV, lived there in the eighteenth century. The Tauscher family bought the brewery, which sits directly behind the hotel, in 1847, and Fritz Tauscher is a seventh-generation brewer. The *Kronen-Brauerei* is the last of 26 breweries that once operated in Tettnang. It produces about 6,000 hectoliters (something more than 5,000 barrels) a year, about 60 percent of that sold in bottles labeled "Tettnanger."

Tauscher is one of nine brewers in a group its members call *Brauer mit Leib und Seele* (brewers with body and soul). "All are owners of breweries in the hands of their families," he explained. "The beers are brewed with our hands."

Each room of the hotel has its own name—thus *Bierbrauer, Bärenplatz,* and of course *Hopfen*—decorated to that theme. Although the town may be best known for hops, its population is growing because of high tech businesses and tourism. Old World and New World coexist comfortably. Tauscher, who was born in 1980, has been to the Craft Brewers Conference in the United States and has seen a future full of beers brimming with exotic hop character. "I can imagine I will brew one or two of those beers, but not yet," he said. "The beer drinkers here are not ready for these beers."

He uses only hops grown in the Tettnang region, buying them directly from his neighbors, storing the bales in the lagering rooms eight to nine meters below the brewery yard. He learned about them walking through hop yards with his grandfather. "He'd say, 'It is good' or ask, 'Is it better than last year?'" Tauscher said. "For me (Tettnang Tettnanger) is the best hop for producing hopped beer," he said.

He brews his beers using a process made new again, adding 60 to 70 percent of his hops as he lauters wort into the brewing kettle. German brewers at the beginning of the twentieth century often employed first wort hopping, but today many larger breweries use only hop extracts at the outset of boiling to add bitterness. Tauscher conducts a decoction mash to make each of his beers, and lautering takes 120 to 150 minutes. He makes his first hop addition 20 to 30 minutes after runoff begins, and another five minutes before the onset of boiling. He adds hops shortly before the end of the boil, then again in the whirlpool.

His Pilsner, with 34 to 36 bitterness units, is pleasantly smooth yet has a satisfying bite. He explained that initially he added all his first wort hops (what he calls "ground hopping") in one dose. "I thought the bitterness was not so good," he said. He opened his right hand, put it to his chin and slid it down his throat to his clavicle, tracking the path a beer would take. "It was, I'm not sure how you say it in English, *adstringierend*." No translation was necessary.

Brewers interested in packing as much hop character as they can into beer may add hops in the mash, during runoff, throughout the entire boiling process, and beyond. They would like to measure the impact, most often in terms of International Bitterness Units (IBU), but as Tauscher understands, sometimes it does not work to put a number to the sensory aspects of hop perception.

Alpha Acids and Beta Acids

"Alpha acid" is traded globally as a single commodity but in fact refers to multiple alpha acids that are similar in structure but significantly different. The analogues of interest are humulone, cohumulone, and adhumulone, (pre- and post-humulone occur in small amounts). These are isomerized by heat in solution, most often in boiling wort, and each is transformed into two forms, the result being six iso-alpha acids (*cis*-iso-humulone and *trans*-iso-humulone, *cis*-iso-cohumulone and *trans*-iso-cohumulone, *cis*-iso-adhumulone and *trans*-iso-adhumulone). Alpha acids themselves are not bitter and hardly soluble in solutions such as beer. Iso-alpha acids are intensely bitter—four times more than alpha acids—and much more soluble. In addition to providing bitterness, they stabilize beer foam and inhibit the growth of bacteria.

It wasn't until the 1950s that brewers understood more than one humulone existed, although scientists isolated pure humulone and lupulone (the beta fraction) from hops in the 1800s. By the 1930s researchers established relatively accurate utilization rates for humulone in the brewing process, which led to the effort to determine the amount of iso-humulone in a beer. Finally, in 1953, working to develop a way to measure bitterness, Lloyd Rigby and J. L. Bethune separated the three major alpha acids.

Rigby later linked a higher percentage of cohumulone—often referred to simply as CoH (pronounced co-aitch)—with what he described as a harsher bitterness. The ramifications were pronounced, boosting the demand internationally for hops with relatively low amounts of cohumulone, influencing hop research and breeding, and explaining why most analyses of hop varieties include the percentage of cohumulone but not humulone or adhumulone. In the 1990s many brewers interested in making highly hopped beers with "smooth" bitterness turned to low-cohumulone hops, unaware that hops with a higher percentage of cohumulone, which most importantly result in more iso-cohumulones, are more efficient.

Research in the 1990s that contested Rigby's conclusion went mostly overlooked, but a more recent study at Oregon State received more attention. Panelists at OSU tasted beers individually dosed with pre-isomerized extracts from Topaz (high level of iso-cohumulone, about 52 percent of alpha acids) and Horizon (low level, about 20 percent) hops as well as a variety of compounds, including rhohydro-iso-alpha acids, tetra-iso-alpha acids, and even nonhop related additives. Experienced tasters rated bitterness intensity, harshness, and smoothness. They did not find significant differences between the high and low iso-cohumulone treatments. Panelists also used self-generated descriptors to characterize the samples. They judged the Topaz sample the least medicinal and its bitterness the least lingering. (The tetra was significantly more medicinal, harsh, and lingering than either iso-cohumulone treatment.)[1]

Anecdotally, brewers report switching to hops with lower CoH results in a "smoother" bitterness. Those results may be influenced by expectations, or could be a matter of personal preference, influenced (like aroma perception) by genetics. "The quality of bitterness, that's something different," said Dan Carey at New Glarus Brewing. "It's like Eskimos and words for snow. How many do they have?"

Cohumulone and humulone levels vary between 20 and 50 percent each in different varieties, while adhumulone will be 10 to 15 percent. Various studies report that iso-cohumulone is significantly more efficient (increasing utilization) but that a lower percentage of iso-cohumulone results in better foam. Research into which, if any, of the isomers decompose more quickly in beer, resulting in stale flavor, has not produced consensus.[2] In contrast, the difference between *cis*- and *trans*-isomers are well known and of equal importance to flavor stability.

The ratio between the two in a traditionally hopped beer will be 68 percent *cis* and 32 percent *trans*. The *cis* forms may be perceived as more bitter. More importantly, *trans*-isomers deteriorate much faster. Researchers in Germany determined that approximately 75 percent of the *trans*-iso-alpha acids degraded within the first 12 months in beer stored at 82° F (28° C), but only 15 percent of the *cis*-iso-alpha acids. The results remind brewers who use only conventional hops of the importance of cold storage and selling beer when it is fresh. Those who brew with pre-isomerized extracts, primarily because they result in 55 percent utilization as compared to 30 percent, benefit because those extracts contain a higher percentage of *cis*-isomers (85 to 95 percent) and therefore are more stable.

Beta acids are not soluble, nor do they isomerize during boiling to more soluble compounds. However, some of their oxidation products, such as hulupinic acid, can be very bitter, water soluble, and may be found in finished beer. Therefore, as hops age before they are used in brewing their bittering potential is influenced by various oxidative reactions of alpha and beta acids. Additionally, recent research identified various beta-acid transformation products generated during boiling that in sum may contribute to bitterness.

Bitter Compounds (Measured in parts per million)

Compound	Threshold	Concentration typical lager	Concentration "hop pronounced" U.S. craft beer
Cohumulone	5.5	0.3	10
Humulone	7	0.6	10
Adhumulone	7.6	0.4	2
cis-isocohumulone	2.7	12	15
cis-isohumulone	3.2	10	18
cis-isoadhumulone	2.5	3.4	10
trans-isocohumulone	6.5	5.1	9
trans-isohumulone	6.1	4.2	12
trans-isoadhumulone	4.4	1.5	2
Xanthohumol	2.9	0	2
Isoxanthohumol	4.7	0.5	5

Source: "125th Anniversary Review: The Role of Hops in Brewing," Journal of the Institute of Brewing 117, no. 3, 2011

The Bitterness Drift

In the early 1980s Anheuser-Busch chairman of the board August Busch III ordered that freshly brewed cans of *Budweiser* and *Bud Light* would be cryogenically frozen, so that they could be tasted against each other over time. A quarter-century later, the *Wall Street Journal* reported on its front page about how A-B had reversed a decades-long trend by adding more hops to *Budweiser* and *Bud Light*. The underline for the story read, "Seeking Mass Appeal, Brewer for Years Cut Bitterness; Now Drinkers Want More."

Reporter Sarah Ellison described a scene in which Busch and Doug Muhleman, then A-B's vice president for brewing and technology, had cans from 1982, 1988, 1993, 1998, and 2003 thawed and set before them in the St. Louis corporate tasting room. She wrote, "Muhleman … says the company didn't set out to make the beers less bitter. He calls the change 'creep,' the result of endlessly modifying the beer to allow for change in ingredients, weather, and consumer taste. 'Through continuous feedback, listening to consumers, this is a change over 20, 30, 40 years,' says Mr. Muhleman, gesturing toward the row of Budweiser cans. 'Over time there is a drift.'

"The sample cans demonstrate how 'creep' works. The difference in taste between two beers brewed five years apart is indistinguishable. Yet the difference between the 1982 beer and the 2003 beer is distinct. 'The bones are the same. The same structure,' says Muhleman. Overall, however, 'the beers have gotten a little less bitter.' "[3]

Ellison did not reveal specific levels of measured bitterness but did report that Miller Brewing regularly tested A-B's beers and detected higher levels in *Budweiser* beginning in 2003 and in *Bud Light* starting in 2005, apparently in response to a Miller marketing campaign. After years of lowering the level of bitterness in *Miller Lite*, the brewery began raising it in 2001, conducted public taste tests, and used advertisements to pointedly attack what it characterized as a lack of flavor in *Bud Light*. *Miller Lite* shipments, which had long been in decline, grew 13.5 percent in 2004 and 2.1 percent the following year. In one TV spot called "Epidemic" *Bud Light* drinkers ran through the streets shouting, "I can't taste my beer."

What Muhleman referred to as "creep" began well before A-B started freezing cans of beer. Joe Owades, credited with developing the first "light beer" and otherwise well known in the beer industry, told *The New York Times* in 1982 the level of bitterness in beer the previous 10 years had declined about 20 percent. He estimated the bitterness of *Budweiser* at 20 units in 1946 and 17 by the 1970s. Hop usage and bitterness levels likely moved in tandem until the mid-twentieth century, when efficiencies allowed brewers to use fewer hops and maintain the same level of bitterness. Greater efficiency does not, however, totally account for changes in U.S. hop consumption, from 0.65 pounds per barrel to 0.43 in 1950, 0.33 in 1960, 0.23 in 1970, and in 2011 about half of that (when craft breweries are not included).

In order to better track bitterness levels, in 2006 the Barth-Haas Group began to conduct annual analyses of brands from around the world. They measured iso-alpha acids (milligrams per liter), which broadly correspond to International Bitterness Units. In 2009, 11 U.S. lagers averaged 7.6 milligrams per liter, compared to earlier reports that bitterness units were still around 20 in 1980 and 12 by the late 1990s. U.S. lagers, South American lagers, and Chinese beers contained the lowest levels of iso-alpha acids (7 to 9 mg/L).

Iso-alpha acids make the largest bittering contribution to most beers, and the importance of other elements varies greatly. Obviously, highly roasted malts add bitterness, just as they do to coffee. Beers brewed with calcium sulfate are known for a "crisper" hoppy character, while those high in calcium carbonate exhibit a coarser bitterness. Lower temperatures suppress perception of bitterness, so that may rise as a beer warms. The level of polyphenols affects the perception of bitterness.

Bitterness signals toxic danger to most mammals, but recent research contradicts the assumption that human aversion to bitterness is innate. A study including newborns up to six days old and older infants found limited rejection of bitter taste by the newborns, while the older infants (two weeks to six months) consistently rejected bitterness. The authors concluded this suggested an early developmental change in bitter taste reception.[4] That an aversion to bitterness might be acquired helps explain why the mirror reaction, a taste for bitterness, can be learned.

Other research indicates genetics play a major role in determining why one beer drinker may perceive bitterness differently than the next. "Just like some people are color blind, some people are taste blind and simply can't taste bitter things that others can," said John Hayes from Penn State's College of Agricultural Sciences. "It turns out that different bitter foods act through different receptors, and people can be high or low responders for one but not another." Details of a collaborative study, published in 2011 in *Chemical Senses,* provide an explanation for differences reported by other scientists; for instance, that a subject highly sensitive to one bitter compound may be insensitive to another.[5]

Linda Bartoshuk, a physiological psychologist at the University of Florida, coined the phrase "supertaster" in 1991 to refer to people who reported a powerful bitter taste when a chemical called propylthiouracil (PROP) was placed on their tongues. The PROP receptor is one of at least 25 encoded by the TAS2R gene family that respond to various bitter compounds. Now scientists count papillae, tiny structures that house the taste buds, to classify people as supertasters, tasters, and nontasters. Bartoshuk suggested that tasters make up about half the population, with supertasters and nontasters divided equally.

Embracing Bitterness

Belgian brewer and part-time philosopher Yvan de Baets makes a particularly passionate argument about the essential role of bitterness in beer. De Baets, co-founder of Brasserie de la Senne in Belgium, says his appreciation of bitterness began with seeing the pleasure his parents took from food. That "opened the doors to complex flavors, giving me the ability to appreciate them—and that includes bitter tastes," he said. "It made me then want to understand the importance of bitterness for us humans, and the love-hate relationship we have with it." He elaborated via email:

"There are two kinds of tastes in nature: the animal- and humanlike ones. The animal-like ones (we're talking about mammals), are sweetness and fattiness: when an animal is facing those flavors, he directly gets two pieces of information very important for him to survive. He knows he will be provided energy (the reason he wants to eat), but also he knows the stuff he's attracted by doesn't contain poison: it's safe; he can eat or drink it. Therefore, the animal instinct pushes him to sweetness and fattiness. On the contrary, (humans) will reject bitterness. In nature, this taste is a powerful signal of danger, meaning: 'This thing might be poisonous, don't touch it!' Indeed, most plants showing bitterness are dangerous. Nature is, of course, well made. And we are still animals: Our instinct also tells us to rush to sweetness and fattiness.

"But there is at least one difference between us and the animals: the passage to culture. About food, it means that since the origin of humanity, by every population, on every continent, our ancestors have made experiments with all they could find to eat and drink. Step by step, they could see that something bitter is not always bad or dangerous. Sometimes, it can be beneficial to health (a lot of natural cures are made of bitter vegetal matter), but also, and even above all, it can possibly provide us a lot of pleasure—the best of all motivations! It is indeed now proved by brain imagery techniques that the people able to appreciate bitterness get, in comparison, much more pleasure than what the people only able to like sweetness would get from sweet flavors. This is how a *human culture of taste* was born, much richer than the animality of taste.

"So, man is the only animal able to like bitterness, thanks to evolution and culture. It is an acquired taste that needs some efforts and education. It takes time, and some people will never get it. This doesn't please the agro-food industry, which wants to sell massively and quickly their mass-produced, standardized stuffs. To achieve this goal, they are using our regressive instincts on purpose by adding sugar and fat in their junk food. They are then sure to trap the customer in a sort of psychological net he cannot resist. They know instincts are irrepressible.

"I see the bitter beers we make as liquid communication that talks to the people's intelligence, and delivers them from the 'manipulations by the stomach' the agro-food industry is using. By promoting bitter beers, something that had almost been lost forever some decades ago, craft brewers help the human culture of taste to be reborn and to get stronger. They show respect not only for themselves, but also for the people who drink their delightful beers. And they do all this by making something that is a never-ending source of pleasure for their customers. Bitter is definitely better."

Show Me Your IBUs

Eleven American lagers tested by the Barth-Haas Group in 2009 contained an average of 8 IBU. For the sake of comparison, here are the numbers for 10 other beers, for the most part as reported by the breweries where they are made.

Blue Moon White	18	Pilsner Urquell	40
Heineken	21	Samuel Adams Boston Lager	30
New Belgium Fat Tire	19	Shiner Bock	13
Orval	38	Sierra Nevada Pale Ale	37
Paulaner Hefe-Weissbier	12	Stone Brewing IPA	77

Researchers continue to sort out how the brain identifies thousands of bitter compounds with a limited number of receptors. Experiments in Germany determined hop-derived substances activate three specific bitter receptors. The study revealed that the receptors may be broadly tuned and activated by multiple chemically different compounds, that they certainly react differently to the various isomers, and that there is redundancy within the bitter taste system (the same compounds may activate different receptors).

Taste receptor cells within the mouth itself were less sensitive to bitterness than receptor cells in a laboratory test tube. It appeared some of the bitter substances were absorbed by the mucous membrane of the oral cavity and salivary proteins, reducing the effective concentration of the bitter substances that activate taste receptors in the oral cavity.[6] "Clearly, the identified taste receptors, together with adsorption phenomena in the oral cavity, are responsible for the perceived bitterness of beer," Professor Thomas Hofmann of the Technical University at Munich said for a press release that accompanied the study.

They also determined bitterness perception does not increase linearly and does not continue to increase at all above a certain level of intensity— depending, of course, on the individual but approximately 50 milligrams per liter iso-alpha acids.

Understanding IBU and Calculating Utilization

Mikkel Borg Bjergsø, a "gypsy brewer" based in Denmark who makes beer in facilities around the world and sells them under the Mikkeller label, produces one he calls *1000 IBU*. He does not claim a lab would measure 1,000 IBUs in the beer, but simply states that those are the calculated bitterness units. To brew a 10-hectoliter (8.5-barrel) batch he adds 18 cans containing 400 grams (about 14 ounces) of hop extract with 53 percent alpha acids. He also adds "normal" doses of hops for aroma, then 10 kilograms (one per hectoliter) as a dry hop.

White Labs in San Diego, which offers a full spectrum of lab services for small brewers, tested a bottle and determined the beer contained 140 IBUs, a number higher than any reported in any other packaged beer. In contrast, the laboratory at the brewing school at the Catholic University at Leuven, one of the world's most prominent research facilities, measured 96 bitterness units.

Results illustrate not only the vast differences between using a formula to predict how many bitterness units a beer will contain and measuring those units but also the challenge for measuring IBUs in highly hopped beers, particularly those that are heavily dry hopped. The Shellhammer Lab at Oregon State University has suggested that another method, called SBU because it uses solid phase extraction, might be more accurate and useful, but before we look at that let's take a closer look at IBU.

Calculating IBUs

Beers entered in Category 23 of homebrew competitions most often come with a list of ingredients to help explain why an entry should be judged as a specialty beer rather than with those in a particular style, such as India pale ale. At an event early in 2012, one judge read a description of the next beer that included dark malts, "Columbus and Simcoe hops, and 75 IBU."

"Is that Tinseth, Rager, or Garetz?" another asked, tongue firmly in cheek. He understood full well that the number was simply a calculation, perhaps much different than what would be measured in a laboratory. Glenn Tinseth, Jackie Rager, Mark Garetz, and Ray Daniels all wrote formulas to calculate International Bitterness Units, the key differences being how they approach utilization.[1]

Most computer recipe software available today gives a brewer a choice of those formulas and further allows users to pick between them and to make modifications that better suit their system. They produce different results, but any of them may be used to make beer that has more consistent bitterness, which is what the IBU was designed for.

The Tinseth formulas are probably the most widely used, in part because they are available at *www.realbeer.com/hops* and are used by professional brewers at *www.probrewer.com*. Tinseth assembled his formulas in the mid-1990s while studying for a Ph.D. in chemistry at Oregon State University. A homebrewer, he ground hops in the USDA laboratory in return for access to its testing equipment.

He began by collecting utilization information from the professional brewing literature at OSU, including papers from large breweries around the world. He also used data from the pilot brewery at OSU and that small breweries provided. After he established the utilization curve, he measured its accuracy by brewing small batches and testing them in the USDA lab.

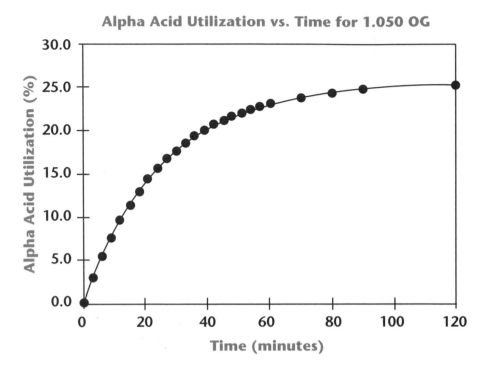

Alpha Acid Utilization vs. Time for 1.050 OG

Glenn Tinseth's utilization chart has served homebrewers well for nearly 20 years, although results will vary in every brewery.

"What's changed for me is how little I would worry about this stuff today," he wrote in an email. "Figure out how your system works, get a few base recipes down that taste the way you want, pay attention to your core ingredients, brew great beer. Consistent brewing practices are far more important than which utilization formula you use. If you are really curious about IBUs, send a sample for testing. Otherwise, don't stress over it unless it makes you happy."

[1] Michael Hall compared them in detail in "What's Your IBU?" *Zymurgy*, Special 1997, 54-67.

Although many breweries who want to signal they use large amounts of hops in their beers use the acronym IBU as a marketing tool, the International Bitterness Unit was created to help brewers make beer with a consistent level of bitterness. It is determined by acidifying and extracting a sample of beer with iso-octane, then taking an absorbance reading at a specific wavelength with ultraviolet light. IBU equals the absorbance times 50. The procedure must be performed in a lab, and the result is an absolute, a single number that is generally misunderstood.

The IBU does not simply measure iso-alpha acids. At the time the formula was developed in the 1950s and 1960s most breweries used hops that were not nearly as fresh as those used today. A certain percentage of their bitterness came from oxidation products, and measuring iso-alpha acids alone did not accurately reflect that. Scientists on both sides of the Atlantic created methods to calculate a number that expressed overall bitterness. The American Society of Brewing Chemists and the European Brewery Convention eventually compromised on the equation that is used today.

The method adjusts the sum of iso-alpha acids and non-iso hop material by a factor of five-sevenths, based on the assumption that five-sevenths of the bitterness of an average beer in the 1960s resulted from iso-alpha acids and the rest from non-iso hop material. This has its shortcomings today, as Val Peacock explained at the 2007 International Brewers Symposium:

"IBU should correspond roughly to actual IAA (iso-alpha acid) content of the beer if, and only if, the beer being tested was made with hops as equally deteriorated as the hops that went into the beers used to calibrate the IBU calculation. This worked fairly well with commercial beers of the 1960s, since more were made with hops which were fairly oxidized by today's standards. But today, with cold storage of hops and the use of hop products that are less susceptible to oxidative degradation before use, the relative amount of non-IAA bitterness in beer is less than the 1960s, and thus the actual IAA content makes up a larger percentage of IBUs. The super-alpha hops used to make much of beer today contain very little beta in relation to alpha as compared to hops 40 years ago. This also reduces the non-IAA contribution to bitterness in today's beers."[7]

Brewers benefit from using the IBU as a tool in formulating recipes and maintaining a specific level of bitterness in regularly brewed beers, while recognizing that it does not perfectly reflect either the quality of bitterness—which will be affected by various reaction processes as well as the composition of the bitter acids—or overall perception of bitterness. It is best measured in a laboratory, but many small breweries settle for estimates, because they either do not have the facilities or do not see the value in sending their beer elsewhere to be calculated accurately.

Homebrewers use formulas that are themselves the result of careful research but nonetheless rely on variables that are nearly impossible to quantify precisely and based on what was known in the 1990s, which has, in some cases, since changed.

"We know what works in our brewery on our equipment," said Matt Brynildson at Firestone Walker, which regularly measures IBUs in its wort and in finished beer. "Craft breweries not running BUs on their beer are making a lot of assumptions."

Even those using a laboratory depend on assumptions. Bitterness units and the amount of iso-alpha acids are equivalent only in the range of 15 to 30 IBUs, and then only when working with relatively fresh hops. Other substances (that do not add bitterness) in beer with no hops will absorb enough light at the specific wavelength to measure two to four IBUs. And, again, recent research in Germany has shown perception of bitterness is not linear and reaches a point of saturation.

A simple approach to calculating IBUs embraces computer software. Many options are available, usually as part of tools for recipe creation, and most allow a brewer to change settings to account for the idiosyncrasies of a particular plant. Occasionally, comparing calculated estimates with actual laboratory analyses and then making adjustments may improve the accuracy of future estimates.

Any discussion about estimating IBUs must begin with utilization, which is influenced by many variables. Change one, the most obvious being boiling time, and utilization changes.

- Form (cones, pellets, extracts, etc.). Hop pellets are approximately 10 to 15 percent more efficient than cones.
- Boiling time and vigor. The relationship between time and utilization is not linear. After 90 minutes, iso-alpha acids break down to unidentified components that are not desirable.
- Kettle geometry. Larger kettles are more efficient, and the difference between a five-gallon homebrew system and even a 10-barrel (310-gallon) commercial brewery is startling.
- Wort gravity. Utilization decreases as wort gravity increases. However, as alcohol and unfermented carbohydrates increase, a beer may support more IBUs.
- Boiling temperature. In an experiment at Oregon State University, less than 10 percent of alpha acids were converted to iso-alpha

acids during a 90-minute boil at 158° F (70° C), while it took only 30 minutes at 248° F (120° C) to achieve 90 percent conversion, possible in a pH 5.2 buffered aqueous solution but not in beer. Water boils at a lower temperature at higher altitudes, lowering utilization.

- The pH and mineral content of the water. Efficiency increases with pH. Of course, higher pH is detrimental to trub formation, protein composition, and yeast nutrition, a reminder that decisions about brewing require constant compromise.
- The composition of the humulones, those higher in cohumulone being more efficient.

Hop utilization equals the quantity of iso-alpha acids found in finished wort or beer, depending on which of these a brewer wants to examine, relative to the quantity of alpha acids added during the boil. The formulas are simple:

$$\text{Hop utilization} = \frac{\text{iso-alpha acids in wort x 100}}{\text{alpha acids added to the wort}}$$

$$\text{IBU} = \frac{\text{quantity x alpha* x utilization* x 0.749}}{\text{beer volume}}$$

* Alpha and utilization are expressed as whole numbers rather than percentages.

To calculate utilization, iso-alpha acids must be measured in a laboratory, preferably by HPLC (high-pressure liquid chromatography, also known as high-performance liquid chromatography), an option being to analyze wort or finished beer. The results, of course, will be unique to the conditions under which the boil is conducted (length of boil, gravity of wort, etc.). An exact measure of hop utilization would also include measuring the alpha acids in the hops (whatever form they might be in) at the time they are added to wort. The number appearing on a package of hops could be the alpha acids at the time of harvest or perhaps after processing. The amount changes during storage, at the merchant or in the brewery, and during shipping. HPLC analysis is more precise than spectrophotometric or conductometric analysis but more expensive.

The IBU estimate is the sum of several calculations, since one must be done separately for each hop addition.

Larry Sidor, an industry veteran who left Deschutes Brewery at the beginning of 2012 to start his own brewery, said that when he was at Olympia Brewing in Washington he would devote what amounted to a day per week blending hops. Olympia used whole cones at the time. "The lab ran IBU after IBU, constantly checking. I had a guy who did nothing but make samples, 10 different lots at a time," Sidor said. The hops were blended to produce beer with consistent IBU levels but also to enhance aroma and flavor.

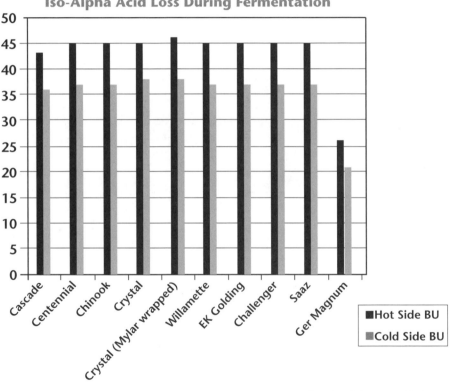

The quantity of iso-alpha acids remaining in finished beer will be further reduced during fermentation and filtration. Preparing beers for a study about hop aroma, Tom Nielsen at Sierra Nevada also measured the IBUs in wort and finished beer. As the graph above illustrates, bitterness levels dropped about 20 percent during fermentation. Brynildson said that iso-alpha acid loss at Firestone Walker is even higher during a vigorous fermentation.

In 2009 Stone Brewing in California made a very strong, highly hopped beer for its *13th Anniversary Ale*. Prior to fermentation the wort measured 130 IBUs, but after beer was bottled it contained 100 IBUs. "IBUs drop during fermentation, because the pH of the liquid drops from about 5.3 to about 4.5," Stone head brewer Mitch Steele said. "This reduces the solubility of the iso-alpha acids, so some bitterness solidifies and drops out and/or gets absorbed by yeast."[8] On average brewers can expect that about 50 percent of iso-alpha acids will be lost in the brewhouse and another 20 percent during fermentation and packaging.

The SBU method under study in 2012 uses solid phase extraction, rather than a solvent, to collect compounds that can be measured using a spectrophotometer. A simple formula returns an SBU number. The first study conducted by OSU in 2011 determined that SBU correlated more closely to iso-alpha acid content as measured by HPLC than did IBU. There was also a strong correlation between sensory bitterness and SBU.[9] The Shellhammer lab planned a second international ring study in the summer and fall of 2012.

Ready, Set, Start Adding Hops

Deschutes Brewery in Oregon describes *Hop Henge Experimental IPA* as its "annual exercise in IBU escalation." In 2008 Deschutes brewers made a batch with 243 *calculated* IBUs, which the lab *measured* at 117 in the fermenter. The bottled beer contained only 87 IBUs *measured*. The next year the brewers included one pound of Amarillo and Cascade hops per barrel as they milled malted barley for the mash.[10] Brewmaster Larry Sidor never considered trying to measure how many IBUs the mash addition might have added. He understood it would be meaningless.

Little brewers do today to add hop impact hasn't been tried sometime in the past, be it hopping in the mash tun during lautering[11] or adding hops at regular intervals throughout the boil. In 1995 the German brewing magazine *Brauwelt* reported on the "rediscovery of first wort hopping," documenting that many German breweries implemented first wort hopping and some experimented with mash hopping. In fact, in the past brewers in England and Belgium also added hops before boiling.

Jean-Marie Rock, who has been the director of brewing at the Orval Trappist monastery since 1985, said that Belgian breweries discontinued the practice in the 1970s. Rock began brewing in 1972, making lagers first at

Palm Breweries and then for Lamot in Mechelen. When Boulevard Brewing brewmaster Steven Pauwels suggested they collaborate on a recipe, Rock knew immediately he wanted to revive the defunct technique. To make the strong Pilsner, 8 percent alcohol by volume with 30 bitterness units, they added two-thirds of the Czech Saaz hops before the outset of the boil.

Rock was happy with the result. "It has a taste you don't get when you use late hopping," he said. "You get an old taste. That is my opinion." Pauwels, a native of Belgium who went to work in Kansas City, Missouri, in 1999, heard about the practice from other Belgian brewers as he learned his trade. He was told they wanted to keep the beer light in color, and the process allowed them to shorten the boil. "It wasn't until later they found out the hop aromas carried over," he said. "It seems like a contradiction. You'd think you'd get more bitterness and less flavor. It's more subtle, almost crisper. Sometimes with late hopping it can get vegetative."

Brauwelt reported that when two German breweries experimented with first wort hopping in 1995 they found the process resulted in beers with a finer hop aroma. Both breweries made two versions of Pilsner in very similar manners, including yeast pitching rates, brewing water, malt lots, and using Type 45 pellets. At Brewery A the first wort addition of Tettnanger and Saaz hops amounted to 34 percent of the weight. At Brewery B, which used only Tettnanger, it was 53 percent. In both resulting beers the first wort-hopped beer had more IBUs, 39.6 to 37.9 at Brewery A, and 32.8 to 27.2 at Brewery B.

Despite increased bitterness, the tasting panel described the first wort-hopped beers as more pleasant tasting and overwhelmingly preferred them. Gas chromatographic analysis indicated the conventionally hopped beers contained a higher level of hop aroma substances (particularly linalool), but panelists nonetheless described the first wort-hopped beers as having a very fine and rounded hop aroma and rounded hop flavor.

The authors of the study concluded, " ... we recommend that first wort hopping be carried out with at least 30 percent of the total hop addition, using the later aroma additions. As far as the use of hops is concerned, the alpha acid quantity should not be reduced even in the case of an improved bitterness utilization. The results of the tastings showed that the bitterness of the beers is regarded as very good and also as very mild. A reduction of the hop quantity could result in the bitterness being excessively weakened, and the good 'hop flavor impression' could be totally lost."[12]

Lupulin Shift: Fact or Fiction?

I'm Having a Lupulin Threshold Shift

🍃 🍃 🍃 🍃 🍃 🍃 🍃 🍃 🍃 🍃

lupulin threshold shift \lu·pu·lin thresh·old shift\ *n* 1. When a once extraordinarily hoppy beer now seems pedestrian. 2. The phenomenon a person has when craving more bitterness in beer. 3. The long-term exposure to extremely hoppy beers; if excessive or prolonged, a habitual dependence on hops will occur. 4. When a "Double IPA" just is not enough.

www.russianriverbrewing.com

Do beer drinkers build up a resistance to hops? In 2005 Vinnie Cilurzo at Russian River Brewing coined the phrase "Lupulin Shift," and the brewery later decorated various wearables with a complete definition.

Two years later the term came up during a question-and-answer session at the First International Brewers Symposium. Following a presentation detailing results of research related to bitterness quality, an attendee explained the concept of "lupulin shift" to Tom Shellhammer and drew an analogy to spicy food. "When you get used to hot food you have to put in more and more spice to get the same perceived spicy heat; the same analogy applies to beer and bitterness, in my opinion," he said.

Shellhammer replied: "I use the same analogy to describe temporal and qualitative effects of bitterness. For instance, the heat from ginger is different than the heat from chili peppers. But in regard to what you described as lupulin shift, we don't see a shift in how the panelists perform over time."[1]

However, human olfactory psychophysics, the study of how humans perceive odors, indicates that the impact of an aroma may change. Andreas Keller and colleagues at Rockefeller University discovered that the perceived smell of an odor at a given concentration changes over time and depends on prior experience. The phenomenon is called adaptation and is caused by repeated or prolonged exposure to an odor, typically leading to elevated thresholds for that odor. Although this does not completely apply to nonvolatile bitter components, it has been shown that the brain, smelling hoppy aromas, expects a more bitter drinking sensation.[2] Adaptation may result.

[1] Thomas Shellhammer, ed., *Hop Flavor and Aroma: Proceedings of the 1st International Brewers Symposium,* (St. Paul, Minn.: Master Brewers Association of the Americas and American Society of Brewing Chemists), 2009, 180.

[2] A. Keller and L.B. Vosshall, "Human Olfactory Psychophysics." *Current Biology* 14, No. 20 (2004), 877.

Because preboil wort has a higher pH level, and also because isomerization begins at temperatures below boiling, the higher measured bitterness levels aren't surprising, but the overall impression has led to confusion when calculating utilization. Some formulas for calculating bitterness suggest treating the first wort addition as a late addition that results in lower utilization. Those are wrong in terms of estimating iso-alpha acids, although they may return a result that more closely approximates the overall hoppy impression.

First wort additions are marginally more efficient than boiling for 90 minutes. Additions to the mash tun, referred to as mash hopping, are not. Little or no isomerization occurs in the tun, and hop material will be left with the spent grain. Some hop material passes into the kettle and will be converted into iso-alpha acids at the same rate as first wort additions. MillerCoors Brewing has run tests in its pilot brewery using pellets, and an addition of 50 parts per million of alpha acids to the mash resulted in beers with between 10 and 15 IBUs for a 10-barrel batch.

Russian River Brewing brewmaster Vinnie Cilurzo mash hops his Belgian-inspired beers but not the highly hopped beers for which the brewery is also famous. He experimented with mash hopping *Pliny the Elder,* a double IPA, just once. "We didn't see any increase in hop character at all, because the beer is so hoppy to begin with," he said. He uses flowers when he hops the mash and finds they make lautering easier.

First wort-hopped and mash-hopped beers may enhance what has been called "kettle hop flavor," which is more easily noticed in moderately hopped beers. Nonetheless, many brewers think those preboil additions result in subtle notes that will be noticed in even hop-intense beers. Kettle hop flavor is not easily described, and the chemistry resulting in kettle hop flavor has not been fully established. "We knew we didn't know the answer," Peacock said, discussing research that focused on linking essential oils to kettle flavor long before he was even in high school.

Looking beyond the lupulin gland, and compounds that mostly evaporate during a vigorous boil, led to the discovery of glycosidically bound flavor compounds in hops that contribute to the complex aroma and flavor matrix. Glycosides originate from a protective mechanism of plants and consist of two parts, one a carbohydrate molecule and the other a nonsugar component called aglycone. In hops different aroma compounds act as aglycones and are variety dependent. Unlike essential oils some of these glycosides survive the vigorous wort boiling process. Combined, their parts are odorless and nonvolatile (so they cannot be analyzed using gas chromatography), but various yeast strains cause individual cleavage of glycosides, freeing the aromatic component and adding to what is called kettle hop flavor.[13]

Much of the early research related to glycosides was conducted at Miller Brewing, now MillerCoors, which brews many of its beers using only hop extracts. Carbon dioxide extraction separates hops into the lupulin gland fraction and a lupulin-free solid. Miller discovered that a beer made with only CO_2 hop extract, which did not contain glycosides, lacked kettle flavor. "We think true kettle hop flavor is a product of yeast and hops," said Pat Ting, a chemist who began working at Miller in 1978 and retired in 2011. "Sometimes it can be similar to hop oil flavor, but it's not exactly the same."

He paused, looking for the best way to describe how it is different. "Crisper," he said. "People usually cannot describe a hoppy flavor. They associate hop flavor with hop oil content, but that is not what they describe."

Ting explained that this flavor does not result simply from hydrolyzed glycosides but also from the subsequent bioconversion by yeast and perhaps even enzymes and microorganisms in the mouth.

Pattie Aron phrased this into the sort of question that might be posed during a roundtable hop discussion at MillerCoors, knowing no one so far has published an answer. "There is another area to look at: How does consumption affect the flavor of these molecules? So, say they exist as glycosides in the product being consumed, what happens [in a] mouth that is full of enzymes and other chemical-altering components?" she asked.

When Sidor, who helped modernize American pelletizing when he worked at S.S. Steiner, saw the results of the research at Miller, it made perfect sense to him. He had seen green matter left over during production of Type 45 pellets and smelled what was in the air. "When you don't have that flavor (apparently from glycosides) the beer becomes less drinkable," he said. "How many beers do you have that smell great, then you get to the middle and think, 'Wow, where did it go?' "

Among the many hop patents Miller owns are ones for isolating glycosides from the vegetative part of the hop cone and from the hop plant itself. Brewers who use only CO_2 hop extract and downstream products may find they need a "glycoside addition" for kettle hop flavor. That flavor occurs naturally in conventionally hopped beers, but brewers who use cones or pellets still benefit from understanding the role glycosides play. For instance, German researchers found the same aglycones in five different hop varieties they examined (all German), but with distinct differences in the content of the glycosides.[14] As researchers examine more hop varieties and yeast strains it seems likely they discover more differences.

The rules for making hop additions during the boil are about as well defined as those for a knife fight. For example, asked about the value of different additions during the First International Brewers Symposium on Hop Flavor and Aroma in Beer in 2007, Dietmar Kaltner of S.S. Steiner said: "I think you have to differentiate between the new brewhouse systems with 60 or 70 minutes of boiling time or even less and the old-fashioned ones with 90 minutes. Here it makes more sense to divide into three portions. If you have 60 minutes, you better do it in two additions. In case of hop flavor, I would use bitter and aroma at the beginning of boiling and the necessary amount of very fine aroma hops at the end or in the whirlpool. The reason for the combination of aroma and bitter hops at the beginning of boiling is the impact of unspecific bitter substances, which are higher in the aroma hops. These bitter substances, which are positive for the quality of bitterness, need a certain boiling time. With additions at the end of the boil, you can't bring them in solution. But there is no common rule for a hop recipe. Every beer has a different matrix."[15]

In 1897 author W.E. Wright first offered readers more standard advice in *A Handy Book for Brewers,* suggesting three hop additions, then proposed an alternative: "Another way, a somewhat niggling one, but with a good deal to recommend it, is to divide the hops for each copper into fractional parts, say, for argument's sake, into tenths, and then as each tenth of the copper length is got, to add one portion of the hops, save as regards the last tenth, which is to be of the best hops used in the day's brewing, and which is only put into the copper half an hour or so before 'turning out.' "[16]

This would seem to be a precursor of "continuous hopping," a process made notorious by Dogfish Head Craft Brewery in Delaware. Founder Sam Calagione said he took inspiration from a television chef who suggested adding ingredients for a soup in equal increments would result in more integrated flavors. Gas chromatography won't measure how well integrated the hop flavors might be, but *60 Minute IPA* is Dogfish's best selling beer, and demand for *90 Minute IPA* and *120 Minute IPA* exceeds supply.

Calagione rigged up a plastic bucket and vibrating electronic football game he bought at the Salvation Army to add hops in regular intervals for the first five-barrel batch of *60 Minute IPA* brewed at Dogfish's Rehoboth Beach brewpub. When the brewery began making *90 Minute IPA* at its production facility, a brewer would stand over the kettle, continuously tossing in pellets for 90 minutes. A mechanical hopper, dubbed Sir Hops Alot, automated the process, and when Dogfish replaced its 50-barrel brewhouse with a 100-barrel system it added Sofa King Hoppy, a pneumatic cannon connected to the brew kettle with hard piping.

Dogfish also continuously hops *My Antonia,* a beer first made in collaboration with Italian brewery Birra del Borgo. Birra del Borgo does not have a cannon, so when the brewery cast of Rome produces its version a brewer must stand over the kettle for the entire boil, making the incremental additions. "Leo (brewery founder Leonardo di Vincenzo) tells me that when they brew the beer everybody is cursing *that Sam Calagione,*" Calagione said.

Di Vincenzo brews several beers rich in American hops, including *ReAle Extra,* the "error" version of his *ReAle.* He said he once forgot to add hops until he was almost finished brewing a batch of *ReAle,* then dumped in an entire recipe's worth in the final five minutes of the boil. Now he makes all the hop additions for *ReAle Extra* in the final 10 minutes, using three times more hops than in *ReAle.*

American homebrewers gave the technique a name: "hop bursting." This amounts to making all or almost all of the hop additions in the final 20 minutes of the boil and using a larger, sometimes much larger, quantity because of lower utilization. There has been little analysis of beers made in this manner, either technical or from brewery sensory panels, but the impact is similar to additions made in the whirlpool or using a hop back.

Brewing With Fresh Hops

Fresh hop, wet hop, or harvest beers, by whatever name, appeal to brewers and drinkers alike. "We promote the heck out of them (in the Northwest)," said longtime brewer John Harris. Brewers in Oregon and Washington make scores of them, and festivals each weekend celebrate them, but their impact is national. Hopunion, the largest vendor of fresh hops but not the only one, sold 120 "Green Hop" packages in 2011, shipping them UPS next-day air. Additionally, breweries across the country include locally grown and freshly picked hops in their own harvest beers.

The hops must be used within a matter of days, preferably one day, after they are picked, or they will begin to rot. Brewers who do not have access to large amounts of fresh hops sometimes use dried hops, often in pellet form, for bittering, and save wet hops for late additions, even post-fermentation.

Results of studies that track how dramatically essential oils change in the days before hops are picked imply that wet hops, which do not contain the same oxygenated fractions as kilned hops, may produce different odor compounds than those that are dried. Unfortunately, no similar studies about wet hops have been published. "This is not a scientific exploration of brewing," said Ninkasi Brewing co-founder Jamie Floyd. "Where's the economic benefit of analyzing a beer made once a year?"

Vinnie Cilurzo at Russian River Brewing draws an analogy to the differences between using fresh and dried basil. "They both do a great job, but you get more fresh aromas and flavors when the product is wet, and it takes more, as you have to compensate for the water that is still in the hops," he said. "I find more melon and grassy notes in wet hops, grassy almost like a Sauvignon Blanc."

Sierra Nevada Brewing began making the most influential commercial wet hop beer in the United States in 1996, one now called *Northern Hemisphere Harvest Ale*. (See recipe and more details about brewing with fresh hops, p. 266) Brewmaster Steve Dresler credits hop merchant, and hop mentor to many, Gerard Lemmens with telling him about a beer Wadworth & Company brewed initially in 1992, using one of the Golding varieties. Wadworth has made *Malt & Hops* every year since.

The late Michael Jackson described the beer as having the "lightest touch of malty sweetness to start; then a surge of cleansing, refreshing, resiny, almost orange-zest flavors; and, finally, an astonishingly late, long finish of fresh, appetite-arousing bitterness." Three years before Sierra Nevada brewed its first wet hop beer, Jackson wrote that he could think of only one other brewery, "in the far west of the United States," making such a *bière nouvelle* but did not identify the brewery.

Many brewers, of course, aim for hop impact, although not necessarily beers as bitter as those brewed with hops that are dried after they are picked. "In order to taste and feel the hops, you have to put them in the right kind of beer," Harris said. "If the beer gets too bitter, you start losing the nuances of the fresh hops."

Dresler agrees. "I really want the aroma. The freshness is what it's about," he said. "If you try to drive up the bitterness (by adding more hops), you'll start to get those grassy notes, chlorophylly."

Jeremy Marshall at Lagunitas Brewing notices that each year. "When I taste it the first 12 or 24 hours (into fermentation), all I get is chlorophyll," he said. "The first time we made (a wet-hopped beer) it tasted like a cigar. I almost dropped a batch because of the cigar taste. Then it starts to open up, the oils go through."

Although brewers such as Sierra Nevada and Great Divide Brewing in Colorado successfully package wet hopped beers with reasonable shelf lives, most of these beers are served on tap.

"I think they fall apart too fast to put them in the bottle," Harris said. "In a month they are a different beer."

Post-Boil Hopping

Neither hop backs nor whirlpools were originally conceived for the purpose of adding hop aroma. Historically, hop backs were used primarily to strain spent hops and trub from wort before it was cooled prior to fermentation. A traditional hop back designed to use with whole hop cones—most often found in England, including at good-sized regional breweries—has a slotted false bottom. Wort is transferred into the hop back and allowed to settle, creating a filter bed of hops on the false bottom. By adding fresh hops before running wort into the hop back, brewers both solidify the filter bed and add fresh hop aroma.

German manufacturer ROLEC, whose clients include several large American craft brewers, makes a device it calls HOPNIK that works like a hop back, and one called DryHOPNIK for use in dry hopping in the cellar. The first is designed for use with flowers and the second with pellets. Victory Brewing in Pennsylvania was the first brewery to buy a HOPNIK, which it since has replaced with a larger vessel designed in-house and called HopVic. Victory uses the HopVic both as a strainer and to add hop aroma to several brands, including *Prima Pils, HopDevil* (an IPA), *Hop Wallop* (an imperial IPA), and *Headwaters Pale Ale,* by adding hops between the kettle and fermentation tank.

Victory installed the HopVic because the HOPNIK was undersized for the demands the brewery put on it, and Victory wanted faster throughput to enhance hop aroma. In fact, IBUs went up about 10 percent, so Ron Barchet shifted hop additions to later in the boil to reduce the IBUs. The intensity of the flavor and aroma both increased. "The improvements are mostly realized in hoppy beers, but even the less hoppy beers are going through the HopVic quicker, which allows for better whirlpool function," Barchet said.

The HOPNIK system consists of a cylindroconical tank with a double-wall and sieve inside. Hops and wort enter to the inside, the wort passes through the sieve, and the hops remain inside. The wort flow is partly tangential to increase extraction and to avoid blocking of the sieve. "We see a trend of hop backs coming back," said Wolfgang Roth of ROLEC. "The trend started with the U.S. craft industry is swapping over to other parts of the world."

Other breweries using a HOPNIK include Thornbridge Brewery and Meantime Brewing in England and Odell Brewing in Colorado. Meantime uses HOPNIK to dose all its hop additions and expects to get 14 percent utilization during whirlpool and 12 percent with the hop back addition.

No studies compare beers produced using a hop back to other methods intended to highlight hop flavor and aroma, but brewers themselves often comment on a difference. Those include brands from English regional breweries such as Harveys and Black Sheep as well as from American breweries like Odell, Deschutes, Tröegs Brewing, and Victory. In 2011 Deschutes and Boulevard collaborated on a beer that combined the Belgian white style with an American IPA. Each brewery made the beer on its own system. Boulevard's Pauwels immediately noticed the difference when the breweries exchanged test batches. "We use pellets, they use flowers and have the hop back," he said. "They can do a lot more in the brewhouse. We have to do more with dry hopping. It was sure a lesson for me how you can use the brewhouse to get more out of the hops."

A whirlpool is used to remove hop fragments as well as other proteinaceous trub, but also may be employed to add hop character. Brynildson reports that utilization can be as high as 22 percent when fresh pellets are added to the whirlpool—late in the process, to minimize contact time with hot wort, which will be less than an hour—at Firestone Walker. That utilization will diminish in beers with higher specific gravity and more highly hopped. "I think, conservatively, you can expect 15 percent on a 11.5 to 12.5 (Plato) beer," he said. Other breweries, including brewpubs, that sent their beers out for analysis recount similar utilization.

Brynildson said that HPLC analysis showed there are also an appreciable amount of solubilized non-isomerized alpha acids in these beers that contribute to the flavor and affect the IBU analysis. Not surprisingly, the results are similar in dry-hopped beers.

To determine which hopping processes were most effective in generating hop flavor and aroma, all 35 breweries in the Rock Bottom Breweries group made the same beer, an India pale ale, but varied the time and manner of the final hop addition. The exercise provided brewers with a better understanding of the impact of different hopping schemes. Van Havig, a regional brewer for the group at the time, organized and reported on the project.

Each beer received the same significant hop dose for 90 minutes and for 30 minutes. Breweries then followed one of four procedures: 1) added one pound of Amarillo (8.4% alpha) per barrel at the end of the boil, with 50 minutes of post-boil residence; 2) added one pound of Amarillo per barrel at end of boil, 80 minutes post-boil residence; 3) added one-half pound of Amarillo per barrel at the end of boil, with 80 minutes post-boil residence, and one-half pound of Amarillo per barrel as dry hops; 4) dry hopped with one pound of Amarillo per barrel, with no additional kettle addition.

A sensory panel that included 34 experienced tasters later evaluated the beers based on seven characteristics: perceived bitterness, intensity of hop aroma, intensity of hop flavor, malt character, citrus notes, fruity notes, and grassy-vegetal notes. In assessing the results Havig cautioned that these were intensely hoppy beers and that Amarillo is a very distinctive hop. Nonetheless, the results were statistically significant and resulted in several conclusions:

- Longer post-boil residence (procedure 2) resulted in more hop flavor, aroma, and perceived bitterness than shorter. This supported an initial hypothesis. "This is in contrast to a commonly voiced opinion among craft brewers that volatile hop oils are quickly driven out of hot wort, and therefore, wort cooling should happen as quickly as possible after the addition of final hops at or near the end of the boil to preserve the hop flavor and aroma in the wort," Havig wrote.
- Longer post-boil residence resulted in more hop flavor than dry hopping, and that hop flavor is best developed in the kettle.
- There was no apparent relationship between measured bitterness and hop flavor or hop aroma, but significant correlations between perceived bitterness and hop flavor or hop aroma. Havig wrote, "This result also brought into question the usefulness of using IBU as a method of measuring the hop character of very hoppy, IPA-style beers."
- A combination of late hopping and dry hopping (procedure 3) resulted in greater hop aroma than longer late hopping. However, it appeared there was a diminishing return for additional quantities used in dry hopping.[17]

Little wonder dry hopping has become so important at many breweries.

Notes

1. Thomas Shellhammer, ed., *Hop Flavor and Aroma: Proceedings of the 1st International Brewers Symposium* (St. Paul, Minn.: Master Brewers Association of the Americas and American Society of Brewing Chemists, 2009), 175-176.

2. Val Peacock, "Percent Cohumulone in Hops: Effect on Bitterness, Utilization Rate, Foam Enhancement and Rate of Beer Staling,"

presentation at the Master Brewers Association of the Americas Conference, Minneapolis, 2011.

3. Sarah Ellison, "After Making Beer Ever Lighter, Anheuser Faces a New Palate," *Wall Street Journal*, April 26, 2006.

4. H. Kajiura, B.J. Cowart, and G.K. Beauchamp, "Early Developmental Change in Bitter Taste Responses in Human Infants," *Developmental Psychobiology* 25, issue 5 (1992), 375.

5. J. Hayes, M. Wallace, V. Knopik, D. Herbstman, L. Bartoshuk, V. Duffy, "Allelic Variation in TAS2R Bitter Receptor Genes Associates with Variation in Sensations From and Ingestive Behaviors Toward Common Bitter Beverages in Adults," *Chemical Senses* 36, vol. 3 (2011), 317-318.

6. D. Intelmann, C. Batram, C. Kuhn, G. Haseleu, W. Mcyerhof, and T. Hofmann, "Three TAS2R Bitter Taste Receptors Mediate the Psychophysical Responses to Bitter Compounds of Hops (*Humulus lupulus* L.) and Beer," *Chemical Perception* 2 (2009), 131.

7. Shellhammer, ed., 164-165.

8. The recipe provided to brewers who wanted to replicate the *13th Anniversary Ale* called for a total of 5 pounds of hops per barrel, including 1.5 pounds per barrel of Chinook (13% alpha acids) boiled for 90 minutes and one pound per barrel of Chinook at knockout. The beer was dry hopped with 1.5 pounds per barrel for 7 days. Those hops were discarded, and it was dry hopped with 0.5 pounds per barrel for another 7 days.

9. Thomas Shellhammer, "Techniques for Measuring Bitterness in Beer," presentation at the Craft Brewers Conference, San Diego, 2012.

10. When I posted this information on Twitter, @robbiexor responded, "Forget hopping during malting, let's just crossbreed barley with Cascade."

11. Green Flash Brewing in California mashes and sparges a beer it calls *Palate Wrecker* using hopped wort.

12. F. Pries and W. Mitter, "The Re-discovery of First Wort Hopping," *Brauwelt* (1995), 310-311, 313-315.

13. Shellhammer, ed., 35.

14. H. Kollmannsberger, M. Biendl, and S. Nitz, "Occurrence of Glycosidically Bound Flavor Compounds in Hops, Hop Products, and Beer," *Brewing Science* 59 (2006), 84.

15. Shellhammer, ed., 47.

16. W.E. Wright, *A Handy Book for Brewers* (London: Lockwood, 1897), 317-318.

17. Van Havig, "Maximizing Hop Aroma and Flavor Through Process Variables," *Master Brewers Association of the Americas Technical Quarterly* 47, no. 2 (2009), doi:10.1094/TQ-47-2-0623-01, 4-6.

8

Dry Hopping

Scores of methods exist, but the intent remains the same: Aroma impact

Paul Farnsworth was 16 years old in 1961, when he began working as an apprentice at Truman, Hanbury & Buxton's Burton brewery. The brewery dry hopped its premium cask ales, always with two ounces of Goldings in a firkin. "It was a three-man job," Farnsworth said. "One held the copper funnel, one the bag of hops, and one used a stick to jam them in the firkin. If one of them was out, then beer didn't get dry hopped that day. Union rules. The invention of (hop) plugs eliminated a job."

Farnsworth went off to university in 1966 to become part of the first generation of English brewers trained in brewing science rather than through an apprentice system. He studied microbiology, earning a Ph.D. from the University of London in 1973. He saw technology evolve within English breweries —for instance, larger ones started using hop extracts—and arrived in the United States in 1976 to witness more dramatic brewing change. He planned to spend one year doing postdoctoral work as part of an exchange program. He never left.

He visited New Albion Brewing not long after Jack McAuliffe began brewing and Chico in Northern California to see Sierra Nevada Brewing, and wrote a report for *Zymurgy* magazine. In the years since, along with teaching at various universities, he built and equipped 50 breweries and fermentation plants around the world, set up quality control programs for dozens of small breweries, taught brewing, and occasionally brewed himself. In England one of his jobs had been to calculate how much bitterness hops would yield in the next kettle after they were used fresh in a hop back. In the United States he saw hops used less frugally.

"The U.S. changed dry hopping. That wasn't the way (in the past) you got hop presence," he said.

Cascade

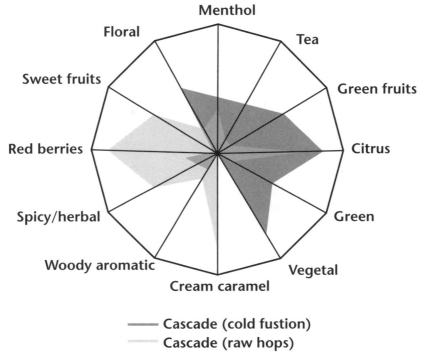

—— Cascade (cold fustion)
—— Cascade (raw hops)

The difference between the lighter shading (raw hops) and the darker (cold infusion) illustrates the potential impact of dry hopping, in this case using Cascade hops. Chart courtesy of Joh. Barth & Sohn.

Dry hopping does not create aromas that perfectly mimic freshly kilned hops—the charts in *The Hop Aroma Compendium* illustrate how a simple infusion in cold water changes perception—but it certainly provides hop presence. As demand for dry-hopped beers increases, brewers need to be concerned about more than finding new ways to pack aroma into their beers. Dry hopping on a large scale is considerably different than adding a few hop cones to a firkin (less than 11 U.S. gallons).

"From a process efficiency standpoint, dry hopping is relatively foolish, but fools who want hops in their beer will do anything to get it," said Saint Louis Brewery director of brewing James Ottolini, who is called "Otto" by just about everybody but one stubborn uncle. Likewise, the brewery is best known by the brand name of its beer, Schlafly, rather than its given name. "We are essentially beer geeks, a combination of fussy artists and pragmatic scientists wrapped up into one," Ottolini said.

Schlafly has experimented with more than a dozen different dry-hopping processes, including the most popular approaches such as pouring hops into the top of the tank or making a slurry, but also with more unusual ones. "It's artistically based on love for beer. Do we like it? Does the customer like it?" Ottolini said. "But the scientist wants to know how it works. We are both romantics and technicians."

The practice of adding hops to casks in England originated before the beginning of the nineteenth century, although the term dry hopping was not introduced until sometime later. In 1796 E. Hughes wrote, "Put some hops in your ale and small beer casks a few days before you want to tap them for use; even those hops that have already been used in brewing will be found serviceable in fining your beer and will not cause it to be too bitter, but will prevent your small beer from becoming sour. Notwithstanding their being used in brewing, they will be found by experience to be very serviceable for the purpose mentioned."[1]

Various primers for brewers in the following years mention the addition of hops to casks to aid fining and preserve the beer, and certainly by 1835 brewers understood that it enhanced aroma. In *A Practical Treatise on Brewing*, William Chadwick suggested reserving a portion of hops for use in the cask. "They will be found to contribute the delightful smell and fine flavour of the hop, much more perfectly than those hops which have undergone a long boiling, and they will equally contribute to the preservation of the beer and prevent any after fretting that might arise," he wrote.[2]

Several American breweries boasted about the quality of their "dry hopped" beers in the years after Prohibition, some listing imported hops, occasionally a variety such as Saaz, and others letting "dry hopped" speak for itself. *Ballantine IPA* stood out among those for several reasons. P. Ballantine and Sons brewed more than 5 million barrels a year in the 1950s, although only a portion of that was IPA. The beer was said to be 7.5% ABV and contain 60 IBUs or sometimes more. Brewers distilled the oils from Bullion hops at the brewery and added them to storage tanks.

After Falstaff Brewing bought Ballantine in 1971, the company closed the Newark brewery and moved production to the Narragansett Brewery in Rhode Island. At first, brewers continued to distill hops and add the oils, but Bill Anderson, who was in charge of brewing at Narragansett, said that soon changed. Instead, they ran Bullion (and later Brewer's Gold and American Yakima, a Cluster-related hop) through a hammer mill before dry hopping. "We ground them to a consistency that was a cross between corn flakes and sawdust," Anderson said.[3] He said the brewery soon dropped the IBUs to 50, then to 45, and reduced the ABV to 6.7%. Pabst bought Falstaff in 1975, moved production to Fort Wayne, Indiana, in 1979, and most agree *Ballantine IPA* never had the same character again.

Today the term dry hopping refers to the addition of hops in the fermentation vessel, in maturation vessels, or in casks. Experiments at the beginning of the twentieth century that included the addition of hops during fermentation apparently were motivated more by economy than by improving aroma and flavor quality. Walter Sykes and Arthur Ling described two patented processes in *The Principles and Practice of Brewing*: "Among the more recent suggestions for effecting economy in the use of hops, the following may be mentioned: H.W. Gates … proposes to hop beer in the fermentation vessel by lowering into it a special cylindrical receptacle having perforated walls, filled with raw hops, and pumping the beer from the bottom of the fermentation vessel into the top part of the hop cylinder so that it percolates through the hops. When it is not desirable to pump the fermenting beer, the hops may be packed in a perforated cylinder fitted up so as to serve as a rouser, which is stirred about in the beer during fermentation. The hops after being thus extracted are said to be suitable for use in the copper. A.J. Murphy … has also devised an apparatus, consisting of a number of perforated tubes or cylinders arranged in a vertical position between perforated top and bottom plates, and around a central rod, for aerating fermenting liquids and for preparing vegetable and other infusions. This apparatus may be used for hopping beer. For this purpose it is charged with hops and alternately raised and lowered in the liquid."[4]

Relatively few breweries today dry hop during primary fermentation. Fuller's in London is one, varying the process from one brand to another, sometimes adding pellets at the start of fermentation and other times at the end. Its brewers may also drop whole hops in muslin bags in maturation tanks during the warm maturation period or put plugs (Type 100 pellets) in casks. For instance, the brewers dose *Chiswick Bitter* with Target in the fermentation vessel, Golding in the maturation tank, and Golding in the cask. Brewing manager Derek Prentice said each method creates different characteristics, ones that are tasted rather than measured in a laboratory.

The Schlafly Tap Room opened its doors in downtown St. Louis in the final days of 1991 and was one of the first brewpubs in the United States to regularly serve cask-conditioned beers. Dan Kopman, one of the co-founders, shipped empty firkins and hop plugs from Scotland, where he was living at the time. Dry hops "added an element of 'oops' to serving casks," Ottolini said. Adding whole leaf plugs resulted in better tasting beer, but it was much harder to serve until the brewers set the firkins on chalks, and later built cradles, so hops didn't end up in the glass.

Schlafly still serves dry-hopped cask beers at the Tap Room and at Bottleworks, the restaurant attached to its production brewery. To dry hop draft beers at the Tap Room, brewers add pellets through the pressure relief valve at the top of a maturation tank. "It's one thing to climb eight feet (on a ladder), and another to put on a harness and go 25 feet in the air," Ottolini said, explaining safety issues make the same process unsuitable at the larger brewery with taller tanks.

Since Bottleworks opened in 2003 a few experiments have included using a hammer mill to pulverize hops before adding them to a maturation tank, converting a hose cleaning trough into a "hop coffin," and swinging beer back and forth between vessels as part of a "pendulum method." Some led more directly than others to two different methods used at Bottleworks, one for whole hops and the other for pellets.

Others simply became part of history. For instance, the first time Schlafly dry hopped in something bigger than a firkin, Taproom brewmaster Stephen Hale stuffed pellets and cones into paint-strainer bags and zip-tied them onto a stainless chain inside a brewpub unitank. When the Bottleworks production facility opened, the brewers began their search for a more efficient method. They experimented with loose whole hops in the bright tank, but discovered when hops floated to the top those in the upper layer remained dry and presumably still full of their essential oils. Later the brewers tried something they called a "reverse French press," using stainless steel screen plates to hold hops down. However, the buoyancy of the hops lifted the screens in places, and dry layers resulted. "Everybody hated it," Ottolini said. "Sanitizing a screen that is six feet in diameter; tedious does not describe it."

After the "hop coffin" at the Saint Louis Brewery proved unsatisfactory it was set to the side and ended up being stolen.

The "hop coffin," a hose trough fitted with a lid, was the first attempt to include an exterior vessel in the process. Brewers laid hops on 16 2-foot-by-1-foot screens, each set at an angle in the trough, then pumped beer from the maturation tank through the hops and back into the maturation tank. However, they discovered rectangles make poor pressure vessels, and the coffin couldn't hold more than five pounds per square inch. Although it didn't work, it was another step toward the method Schlafly uses today, developed about the same time as Sierra Nevada Brewing's "torpedo," although neither brewery knew about the other's experiments.

Schlafly later converted two seven-barrel fermentation vessels into the "hop rockets" used to make its *Dry Hopped American Pale Ale*, fitting the cones with small lauter tun plates and replacing the CIP (clean in place) ball, a part used to clean the tanks, with a slotted tube that extends into the hop bed. Brewers put four mini-bales, totaling 52 pounds of whole hops, in one vessel and circulate beer from a 200-barrel tank through the hop rocket for one week.

The brewery also tested dry hopping the popular beer with pellets, but in-house objections quickly blunted that. "We got an outcry from the people making the beer, 'It's different.' We did a triangle (blind taste) test and proved to ourselves there was no difference," Ottolini said. "Still, it felt right to support some strong-willed and vocal employees. We want the people making the beer to be happy with the way they are making beer."

Schlafly experimented in 2011 with a "pendulum method" to make its *American IPA*, known locally at *AIPA*, literally swinging beer back and forth, using three vessels. Brewers loaded one with hops, purged it with CO_2, and filled it with beer from the fermentation vessel. After two days they transferred the beer into a holding tank to measure the yield, blending it back into the bright tank after the tests. They repeated this process three times for one batch, four times for the next, and five for a third.

AIPA contains 7.2% alcohol by volume, is hopped with Simcoe, Centennial, and Amarillo, and is full of the bright, fruity aromatics that have come to characterize the most popular American IPAs. The brewery released five different batches during the spring and summer. Brewers didn't use the pendulum method for the final batch, instead placing pellets in a fermentation tank, transferring in beer, and circulating it. "(The result) was easily the most intense," said Matt Murphy, who administered the trials. "We thought it might be too much."

That version turned out to be the favorite of the most vocal of St. Louis beer enthusiasts, drawing raves on the discussion board at STL Hops, a local internet site. Ottolini joined the conversation to say the final release "had everything the hops had to give," including some flavors not traditionally considered positive.

"We've got a bias on our taste panel. What we like is traditional hop flavor. We're still trying to figure it out, what we want in hops ... from hops," Ottolini said. As brewers, what they prefer themselves doesn't necessarily come first. "It doesn't matter what we like. This seems to be what the consumer wants," he said, speaking specifically of the last, particularly pungent, batch from 2011, one that contained additional vegetative hop matter.

Not surprisingly, Schlafly moved on to another process in December 2011. The approach is similar to but not totally the same as the slurry methods described in this chapter. Although brewers used a centrifuge to remove hop particles from the beer, Ottolini looked to other food industries for an alternative, and results of planned experiments with an inline hydrocyclone could be of interest. The challenge is keeping the density (blend) of the solids and liquids consistent, because the ratio of oils to vegetative substance is seldom constant, particularly at breweries that dry hop with a combination of Type 45 and Type 90 pellets.

The method is much simpler when Schlafly makes a batch on a smaller scale. Early in 2012 Jared Williamson, a shift brewer, dry hopped a batch of pale ale to be served only at Bottleworks. "After 36 hours of primary fermentation the beer was within 1° Plato of its limit, at which point I dumped yeast off of the cone, dropped (Amarillo) hop pellets through the PRV port on top of the tank, and bunged the tank," he explained. After pressure built to 14 pounds per square inch and the beer was diacetyl negative, he crashed the temperature. "I periodically dumped hops off the cone as they settled out."

When the Shellhammer Lab at Oregon State University began research related to dry hopping and aroma in 2011, graduate student Peter Wolfe surveyed small breweries about their dry hopping practices. He quickly found the only consistent aspect of the way various breweries dry hop beer is that there is no consistent approach. Not how they add hops. Not the temperatures at which they dry hop. Not the amounts they use, the amount of time the hops remain in contact with the beer, if they first remove yeast, if they dry hop a second or even third time.

"I think everybody is still experimenting. You can always fine tune it to get better aroma," said Boulevard brewmaster Steven Pauwels, who moderated a panel discussion about dry hopping at the 2010 Craft Brewers Conference in Chicago. After the conference, Boulevard began dry hopping in two stages, sometimes three, instead of using a single addition. "We let it sit two days, blow it out, put in more," he said.

Peter Bouckaert from New Belgium Brewing in Colorado was one of the speakers in Chicago. "If you'd asked me 10 years ago if I'd be on a dry-hopping panel I think you all would have been laughing," Bouckaert said. Sales of India pale ale, or IPA, grew 41 percent in the United States in 2011, and New Belgium's *Ranger IPA* was one of the reasons why.

In *Brew Like a Monk* Bouckaert offered advice as useful as from any brewer I ever interviewed. "Brewing is a compromise," he said at the time. "You have to take into account so many factors. You can't look at the temperature as a sole factor. It's an interaction. You need to see any beer you create as a holistic thing." Although the variables involved in dry hopping may be considered individually, their effects change based upon other decisions a brewer makes.

"The aroma you get is inherent to your method," Bouckaert said.

The Universal Questions

The easiest way to dry hop with pellets is to open a fermenter and drop them in. Boston Beer Company put a glass port in the side of a storage tank in its pilot brewery in Boston in the early 1990s, and brewers watched the pellets initially floating before breaking up as they dropped. When they took samples from multiple points in the tank, they discovered the hop flavor intensity decreased at higher levels. "You can't take just part of that tank and expect uniformity," said David Grinnell, vice president of brewing.[5]

It is a reminder that some variables, such as hopping rate and temperature, may be changed easily. Others, such as the tanks available for dry hopping, must remain constants. "If I were starting from scratch I'd probably put a catwalk around the tank and drop in hops from the top," Russian River's Vinnie Cilurzo said. He doesn't have that option. Russian River uses a hop cannon at its production brewery and adds hops through the PRV at its nearby brewpub.

What might be called the default configuration, a conical tank in a small brewery or a glass carboy at a home brewery, works well. There are plenty of other factors that can be adjusted.

Form

In the first part of the study at Oregon State, Wolfe dry hopped beers with both pellets and whole hops under cold conditions, 32° F (0° C), then had a sensory panel evaluate them for overall hop impression after one day of dry hopping and after seven. He also analyzed the beers using gas chromatography. Pellet aroma was consistently rated as more intense than whole cone aroma by the panelists. The sum of the aroma peaks recorded on the GC confirmed pellets produced a larger amount of aroma compounds.[6]

Brewers already understood that hop oils and compounds in pellets dissolve more quickly into beer than those in cones, so dry hopping with pellets produces quicker results. Not surprisingly, discussion about the quality of aroma and flavor that result from dry hopping with the different forms is a subset of the larger whole hops versus pellets discussion.

"When I was at Anheuser-Busch in the mid-'90s we did some extensive testing," said Mitch Steele, now head brewer at Stone Brewing Co. in California. "We thought there was some difference back then. We got some floral character off whole hops that was better than we were getting off pellets. I do believe that was variety dependent."

"I believe pelletizing, the process, has improved a lot in the last 15 years. I don't have any issue with using pellets at Stone. I really like the hop character we get. I do think it is different. It's a careful consideration."

Jeremy Marshall, brewmaster at Lagunitas Brewing in California, is perfectly candid in admitting that pellets sometimes exhibit "a characteristic veggie bite people talk about, although they can't say what they mean. We know what they mean by that. We've tasted the difference."

Temperature

Cilurzo didn't have any choice when he first dry hopped beers at Blind Pig Brewing in Southern California. The only tanks he could use for dry hopping were located in a cold box. After going to work at Russian River Brewing in Northern California he initially continued dry hopping cold. He first raised the temperature to one you would find in a cellar (low 50s Fahrenheit, 11-12° C). "From there we kept making it warmer and warmer until we reached (peak) fermentation temperature," he said.

Dry hopping warmer speeds extraction, an important consideration for many breweries trying to keep up with the demand for dry-hopped beers. It increases the danger of extracting unwanted vegetal and fishy odor compounds, but an experienced brewer may gain better control over just what gets extracted.

Brewers produce excellent beers dry hopped at a wide range of temperatures. The brewers on the panel at CBC in Chicago all dry hop "warm" but nonetheless differently.

- Stone Brewing dry hops at 62° F (17° C). "It's yeast based," Steele said. Stone circulates its beer three times during the process, allowing the beer to finish fermentation and create some carbon dioxide before the tank is crash cooled.

- Lagunitas Brewing conducts a diacetyl rest at 70° F (21° C) for its highly flocculent English ale yeast and dry hops at the same temperature. Marshall said most of the yeast drops during that diacetyl rest and is removed before hops are added.
- New Belgium dry hops at 12° C (about 54° F) so that yeast is still active.
- Sierra Nevada begins dry hopping with whole hops in bags at 68° F (20° C), starting the process when nearly fermented beer is 1-1.5 °P above terminal gravity. "It's critical to have that extract left (and yeast) because even when you purge the tank with CO_2 there will be oxygen trapped in the sacks," brewmaster Steve Dresler explained. Brewers lower the temperature during the two weeks that bags remain in the beer.

In contrast, Boston Beer dry hops *Samuel Adams Boston Lager* inline as it is transferred from a fermentation tank to a lagering tank after mixing hops in with cold, de-aerated water to create the slurry (page 226).

As a data point, one dry hop experiment at OSU included very warm extraction, 77° F (25° C), over a short period of time. Wolfe used both cones and pellets agitated during extraction. He gathered information on levels of myrcene, humulene, limolene, linanool, and geraniol from 30 minutes through 24 hours. He discovered extraction of the terpenes (myrcene, humulene, and limonene) peaked between three and six hours in dry-hopped beers and then declined, while the concentrations of terpene alcohols (linalool and geraniol) continued to increase throughout the 24-hour dry-hop extraction. This was one of the first trials in a larger experiment under way in the spring of 2012.

Quantity

Again, from Chicago:

- Stone uses between one-third and 1¼ pounds per barrel for dry hopping, based in part on the starting gravity of a batch. Most beers include one-half to one pound.
- Lagunitas additions average between one-half pound and 1½ pounds per barrel, depending on the brand. "There's a saturation constant," Marshall said. "If you do half a pound per barrel you get more bang for your buck than if you do one or two. More

doesn't give you more." Steele agreed the result is not linear. The experiment by Rock Bottom Breweries (pp. 201-202) also indicated diminishing returns when comparing additions of one-half pound and one pound per barrel.

- "There's a ceiling for your method and variety (of hops)," Bouckaert said. New Belgium found that ceiling at 35 kilograms in 100 hectoliters (comparable to about nine-tenths of a pound per barrel).

Chapter 10 includes more examples.

Residence Time and Number of Additions

Firestone Walker *Union Jack IPA* won gold at the Great American Beer Festival in both 2008 and 2009. Russian River *Blind Pig IPA* medaled in 2007, 2008, and 2009. Firestone Walker brewmaster Matt Brynildson favors short contact time for dry hopping, no more than three days. Cilurzo dry hops *Blind Pig* for eight to ten days. That those two beers stood out from more than 100 entries each time is more important for this discussion than the fact the judges preferred *Union Jack* in 2008 and 2009, although it is relevant. Slightly different approaches result in very popular beers.

Both beers are dry hopped twice. Brynildson dry hops *Union Jack* (p. 249) for three days, removes hops and yeast from the bottom of the fermentation tank, and dry hops it for another three days. Cilurzo adds the second dose to *Blind Pig* five days before transferring. He dry hops *Russian River IPA* once for six to seven days and *Pliny the Elder*, a double India pale ale, for 12 to 14 days, again adding the second dry hop addition five days before transfer.

"I've been doing that on *Elder*, two dry hops, from the beginning, back in '99," Cilurzo said. "There was no research. 'Let's try it and see what happens.' "

He calculates the dry hop addition and then splits it into two. "You get a brighter, more intense flavor," he said. "I think you get a lot out of your dry hops earlier. I believe in the importance of perceived bitterness. Longer gives you more dryness on the palate."

The first round of research at OSU hints at that, although more trials are needed. However, anecdotal evidence supports the possibility. For instance, even though New Belgium dry hops a little cooler than many breweries, slowing extraction, it takes only two days. Lagunitas' Marshall has noticed that "after 24 hours flavors start to jump out. After 72 to 96 hours you are pretty much there. … Of course, you ask 10 different brewers and you'll get 10 different answers."

Lagunitas dry hops only once. "I make fun of brewers who do double dry hopping," he said. "Why? Because I'm lazy. I'm lucky it seems to work for us."

Fermenter Geometry

Many things changed when Bell's Brewing in Michigan commissioned a new brewhouse in 2012. One that didn't was the way Bell's dry hops *Two-Hearted Ale*. "I'm usually the engineering geek," brewery manager John Mallett said, "but I can't break gravity."

Brewers at Bell's add pellets through a manway at the top. "There are critical design features we are really aiming to keep consistent," Mallett said. The fermentation tanks are shallow, as are the cones, maximizing surface area for contact between beer and hops.

"I think everything can change as you change the physical tank. Fermentation, aging, dry hopping," he said.

Studies under way in 2012 in both the United States and Germany may provide greater insight into the impact fermenter geometry has on dry hop character.

Yeast

A majority of breweries separate beer from yeast before dry hopping, particularly yeast that will be repitched. Dresler said Sierra Nevada keeps yeast active not only to consume oxygen trapped in the hop bags, but because otherwise compounds produced only by the interaction of hops and yeast may be missing. "We don't get the same floral estery notes in some other beers if we use the torpedo process (described later) simply cold without yeast contact time," he said.

To illustrate the role yeast may play, Steve Parkes, brewer and owner of the American Brewers Guild, has students fill two growlers. One contains nearly finished beer with yeast present, and the other has no yeast. "Then toss in some Cascade," he said. "Should I dry hop before I filter or after I filter? The result will be different. That kind of experiment is simple."

Varieties

Not surprisingly, many brewers save new "flavor" varieties such as Amarillo, Citra, and Galaxy for late hot-side additions and dry hopping. Hops with distinctively "American" flavors are in fashion, including the so-called "C" hops. A half-dozen German brewers poured American-style IPAs, all featuring American-grown hops, at a festival in Munich in the spring of 2012. It would be foolish to predict what the impact of new varieties such as Mandarina Bavaria in Germany or elsewhere in Europe might be. Additional considerations in choosing varieties for dry hopping include total oil content, particularly the percentage of myrcene, and presence of geraniol.

The popularity of Sierra Nevada's *Torpedo Extra IPA* quickly increased awareness of Citra when the hop was new, but Dresler cautions, "You can overbrew with it." The recipe includes Magnum at the beginning of the boil, then a combination of Magnum and Crystal 10 minutes from the end. Each torpedo is packed with Magnum, Crystal, and a "hint" of Citra for dry hopping.

The Slurry Method

Many of the fermentation tanks at New Belgium Brewing are quite tall, with only the lower portion indoors. To add hops through a valve on top brewers had to climb stairs outdoors and a ladder. Safety was an issue. Bouckaert said bees could be as unpleasant in summer as ice was in winter. The brewery experimented with a variety of other processes before settling on a slurry method similar to one used by breweries both large and small.

New Belgium had a 30-hectoliter (approximately 25-barrel) mixing tank specially built. It has a stirring arm and a blending pump. Brewers cool beer to 12° C (about 54° F) after fermentation and a diacetyl rest, centrifuge it, and hook up the mixing tank. They blend hops with 20° C (68° F) de-aerated water. The slurry is ready after one hour and is transferred into the beer.

They drain off spent hops every 12 hours, transfer beer after 48 hours, centrifuge again for hop removal, and chill it to -1° C (30° F).

Stone Brewing added a similar 30-barrel mixing tank in 2011, also fabricated at Paul Mueller Company in Missouri. It became the primary slurry tank, although Stone still uses the old 3.4-barrel grundy in one fermentation cellar.

"I had no idea these guys were doing this much dry hopping," said Steele, who went to work at Stone in 2006. Stone brewed 69,000 barrels in 2007 and almost 150,000 in 2011, about 60 percent of that dry hopped.

"It was a very sound method. We liked the results," Steele said. "We realized as we continued to grow this wasn't the best way to do this. It is a very labor-intensive method. When you are doing a dry hop every day that's not reasonable."

Brewers set up a circulation loop that starts from the racking port of a 400-barrel tank (effectively holding 360 barrels), runs into the grundy, then out of the bottom of the grundy and back into the bottom of the fermentation vessel. Previously, a brewer would put a funnel into the top of the grundy and begin adding hops, between 100 and 400 pounds. It took half a day to dry hop *Ruination*, a double IPA.

The new system operates much as at New Belgium, although the other variables (hopping rates, temperature, and residence time) are all different. Hops are added all at once to the 30-barrel tank, agitated to create a slurry, and then transferred into the fermentation vessel. It is simpler, less labor intensive, takes less time, and had the advantage of being a closed system, reducing microbial and oxidation risks.[7]

O'Fallon Brewing in Missouri uses a simpler slurry method on a much smaller scale. Head brewer Brian Owens adds pellets and warm water to a one-half-barrel yeast brink, blending in one pound of hops for each gallon of water. He blows the mixture in through a standpipe, dispersing it into the fermentation vessel. He adds 11 pounds of hops to a 15-barrel vessel (effective size, 14 barrels), and doses 30-barrel tanks twice. O'Fallon's *5 Day IPA* spends five days on dry hops at 68° F (20° C).

Hop Cannon

Of course a pneumatic device used to disperse distinctly American hops into hopcentric beers would be called a "hop cannon." Dogfish Head Craft Brewery calls its version "Me So Hoppy." Russian River Brewing and Firestone Walker Brewing both use them. Marshall provided a complete overview on building and operating one during the panel session at CBC in Chicago.

Hop Cannon Primer

Tank considerations

- The receiving tank must have at least a 2-inch vent or blowoff port to blow the hops into.
- Long-radius 90° elbows are preferred to standard short-radius 90° elbows.
- There is a benefit to have an oversized vent pipe that reduces down at the valve.
- Venting off of the delivery gas can be achieved through the sprayball piping.

Execution and delivery

- After vessel outfitted with fittings, perform a pressure test.
- Rinse, passivate, and CIP vessel.
- Purge vessel with CO_2 until 100 percent dry (moisture is a huge enemy of the pneumatic method).
- Pour hops under an active CO_2 blanket.
- Seal and purge vessel.
- Connect delivery hose to the vent port of the tank, and send several blasts of gas along through the bypass piping. Tank receiving hops must be vented off continuously.
- Slowly open the bottom outlet to introduce the hop pellets.
- As the hopper becomes less full, the bottom outlet can be opened more.
- It is better to use a higher number of smaller shots at first, and as experience is gained, increase the size gradually.
- Lagunitas can deliver 88 pounds in seven to eight shots (at 80 psi).
- 10- to 12.5-pound shots occur in 3 seconds.
- A pleasant, characteristic "rattling" sound occurs with each shot.
- Most of the time is spent waiting on repressurizing of the vessel.
- An experienced operator can deliver 88 pounds in 30 minutes.

Storage of vessel

- Best to store the dry-hop vessel under CO_2 pressure, which will prevent oils from oxidizing and ingress of microbes.
- Vessel must be washed and sanitized periodically on a calendar basis.

Courtesy of Jeremy Marshall, Lagunitas Brewing Company

Lagunitas decided to try the method after watching a grain delivery, which uses a pneumatic device to propel grain into a tall silo. The cannon is not large. It holds about two 44-pound boxes of pellets so must be reloaded and repurged multiple times to dry hop larger tanks. It proved scalable as Lagunitas added 500-barrel fermentation tanks and its production grew from 72,000 barrels in 2009 to 106,000 in 2010 and 165,000 in 2011. Lagunitas dry hops 90 percent of its beers.

"It's worked with taller and taller tanks," Marshall said. "It's essentially a sparging process, the hops spreading out and hitting the head space. Three hundred to five hundred barrels, we will get the same character. If anything, more aroma, more extraction. The only thing different is the height. The hop particles rain down."

Marshall listed several advantages to the system: safety, relative ease of execution, hygenics, minimal space requirements, and (as important as any of them) improved beer stability. Lagunitas measured 20 to 30-plus parts per billion of dissolved oxygen (DO) in its beer, even with a CO_2 blanket, when hops were added from the top. The level of DO dropped to 4-5 ppb with the hop cannon.

On the negative side, the method works only with pellets, and better with smaller pellets, requires an initial capital expenditure, uses a considerable amount of CO_2, and requires operator training.

Despite the colorful name, Brynildson points out, the cannon does not require high pressure delivery that would make it potentially dangerous. "It is about flow, not pressure," he said.

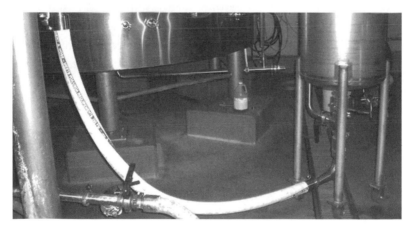

Hop cannon at Russian River Brewing. Courtesy of Russian River Brewing.

Torpedo

Sierra Nevada Brewing uses two different methods to dry hop with cones. One is to attach eight-pound bags of hops, which will be more than four times heavier after absorbing beer, to rings that have been welded to the sides of tanks. *Celebration, BigFoot,* and other small-run beers will soak for up to two weeks. They are then transferred into another tank. "That was really starting to tie up my infrastructure, particularly when we were going to start doing a new year-round dry hop beer," Dresler said.

Sierra Nevada founder Ken Grossman said he had started thinking about a system that forced beer through a bed of hops back in the mid-1990s. That turned out 10 years later to be what Sierra Nevada calls a torpedo. It looks like a torpedo turned on its side. The brewery soon needed more than a dozen to keep up with demand for *Torpedo Extra IPA* and retrofitted even its 800-barrel tanks to work with a series of torpedoes linked together (color plate 8).

A torpedo can hold up to 80 pounds of hops, although Dresler said he prefers the results when it isn't packed tight. It is purged with CO_2, then beer is circulated from the bottom of the cone to the bottom of the torpedo, up through the torpedo and back into the tank. The beer, freshly dosed with hop oils, passes through a tube within a tube (called a periscope) so that it returns higher into the tank. Otherwise, beer in the bottom of the tank may become saturated and won't retain any more hop oil.

The concept is brilliantly simple, although the execution is a bit more complicated. Dresler said results vary dramatically based upon temperature and flow rates. *Torpedo Extra IPA* circulates for five days, beginning at 68° F (20° C) and finishing cold, extracting all the oils Sierra Nevada wants out of the hops much more quickly than with the passive bag system.

Speaking at the 2008 Craft Brewers Conference, Nielsen emphasized the effects of saturation and oil stratification. As a tank is emptied, the dry hop bags become exposed and start dripping oil-saturated beer on to the top layer of the draining beer. The final 2 to 3 percent of the beer being transferred out of the tank contains more than twice as much hop oil due to the accumulation of the "drippings." He drew a laugh when he suggested that last 2 percent could be sold as a select beer.

Dry Hopping in a Glass

Sam Calagione of Dogfish Head Craft Brewery said he was thinking about "real time hopping" when he got the idea for a device he calls Randall the Enamel Animal, otherwise known as an organoleptic hop transducer module. Basically, it infuses beer with fresh hop flavor as the beer passes through a filter full of hop cones between a beer keg and glass.

Calagione first unveiled the device, appropriately enough, for the 2004 "Lupulin Slam" in Washington, D.C., an event at which brewers from the East and West Coasts matched their beers, with consumers voting on which they liked best. Dogfish Head *120 Minute IPA,* served through the Randall, won.

In the years that followed, Dogfish Head built hundreds of Randalls for other breweries, and eventually posted plans for making one on its website. In 2012 Dogfish introduced Randall Jr., a smaller version that did not require a beer tap. It sold out almost immediately. Instructions suggested drinkers could make their own with a tea infuser or French press.

In fact, Bull & Bush Pub & Brewery in Denver began something similar a few months before Dogfish Head introduced Randall Jr. Customers simply start with any house beer, then pick from one of five hop varieties grown in Colorado. A server adds the crushed cones to what Bull & Bush called the Hop Inciter 3000, basically a French press, for however long the customer wants.

Nobody at Boulevard Brewing would laugh. The brewery has fresh Chinook hops flown in each year to make its seasonal *Nutcracker Ale,* dry hopping it with those wet hops. Because the brewery ran at full capacity for several years, tanks turned over quickly, and as soon as *Nutcracker* was ready to package, workers emptied the tanks, emptied the bags, and left the hops to dry.

That changed when the brewery added capacity. "About six years ago I came in and everybody was running around with pint glasses, saying, 'My God, this is amazing,' " Pauwels said, shortly after brewing *Nutcracker* in 2011. They had discovered what the liquid dripping out of the bags tasted like.

The result was *Nut Sack,* which is reserved for special events and occasionally served in the brewery tasting room.

"What's amazing is how it changes from year to year," Pauwels said. "Some years, honestly, it's not good, too grassy. It's all about the quality of the hops, how good the year was. Very volatile."

Notes

1. E. Hughes, *A Treatise on the Brewing of Beer* (Uxbridge, England: E. Hughes, 1796), 30-31.

2. William Chadwick, *A Practical Treatise on Brewing* (London: Whitaker & Co., 1835), 56.

3. Gregg Glaser, "The Late, Great Ballantine," *Modern Brewery Age,* March 2000.

4. W. Sykes, A. Ling, *The Principles and Practice of Brewing* (London, Charles Griffin & Co., 1907), 522-523.

5. In an effort to homogenize hop flavor/aroma intensity within a tank, some breweries use CO_2 to keep hops in, or push them back into, suspension.

6. T. Shellhammer, D. Sharp, and P. Wolfe, "Oregon State University Hop Research Projects," presentation at Craft Brewers Conference, San Diego, 2012.

7. The ROLEC DryHOPNIK is basically a slurry system, using a shear pump to wet mill hops, which are added to a dosing tank that feeds the milling chamber. It is a closed system and portable, designed to pump more than 500 pounds of hops in an hour.

9

The Good, the Bad, and the Skunky

Taking responsibility for hop quality

The Boston Beer Company dry hops *Samuel Adams Boston Lager* with Hallertau Mittelfrüh added as a slurry, completing a cold chain that begins more than 4,000 miles away on a farm in Bavaria. "Those hops are treated reverently," said David Grinnell, vice president of brewing.

Getting the most out of hops requires a certain investment, although few breweries have the resources to baby them quite like Boston Beer does. Larry Sidor put it bluntly. "The biggest single thing mismanaged in most breweries is hops," he said. Brewers must understand the potential negative attributes of hops as well as the positive ones this book has focused on.

"We're a perfect size. Big enough (to support) more difficult practices that are unique for our beer. Small enough to be passionate about hops," Boston Beer founder Jim Koch said. "When we were littler, we were passionate about hops but we didn't have the same influence."

Grinnell explained just what that means.

- BBC helped fund a study that investigated the results of picking at later dates, about the impact of dry hopping with those hops and the resulting flavor stability. Boston Beer now buys hops that are harvested up to seven days later than what was considered the ideal date. They contain between 0.5 and 1.5 percent more alpha acids and correspondingly higher levels of essential oils.

BBC considers this a 20 percent increase in "total brewing value." BBC can use the later-picked lots in dry hopping for more intense aroma or blend them in with the entire production to boost alpha acids and oils.

- Better drying practices. "Probably half (of flaws) are from bad kilning. When we started, in the '80s, most farmers did not have temperature probes," Koch said. "We work with the dealers. They can take our requirements and standards to the farmers."
- Farmers keep bales cold before delivering them to a pelletizing facility, where they continue to be stored cold. "In 2000 we still saw bales sitting out on farms for months, sometimes until springtime," Grinnell said.
- Workers at the plant in Germany that pelletizes hops for Samuel Adams beers use a custom process, creating Type 90 pellets on a Type 45 line. When they otherwise produce Type 45 pellets, those are concentrated, including only enriched hop powder and not much of the vegetative matter that is part of a Type 90 pellet. They mill the *Samuel Adams* pellets at the same super-cold temperature as Type 45s, at about -31° F (-35° C), but retain almost all the hop material. The process leaves more lupulin glands intact, removes some coarse matter, and results in a finer particle size, which disperses better in the dry hop slurry and separates more easily in centrifuging.

Simply using Type 45 pellets would reduce the space needed for storage and deliver more lupulin more efficiently. "Concentrated lupulin is not unattractive, but it's not the whole hop. We're committed to the whole hop," Grinnell said. "Somewhere in your brain you know (that) will have value."

- Pellets are kept cold before shipping, shipped in refrigerated containers, then stored cold after arriving in the United States.
- A dry hop slurry is introduced inline as *Boston Lager* is transferred from a fermentation tank to a lagering tank. Hops are gently mixed with cold, de-aerated water to create the slurry, and Grinnell emphasized the importance of managing oxygen pickup and purging oxygen prior to packaging. Boston Beer originally dry hopped *Boston Lager* by simply adding pellets to a tank. Initial experiments with the slurry method included using warm water.

"We discovered we got an aroma we really didn't want," Grinnell said. "(Now) we feel we get the fresh hop character we smell at hop selection."

"I have found nothing drives up aroma like attention to practical, unsexy details," he said. He estimated the process results in a *Boston Lager* with 15 to 20 percent more hop intensity. "You can use less hops for the same intensity or get more impact," he said. "I'll take more impact."

He and Koch have been selecting hops together for more than 20 years, each harvest assessing more than 500 lots in Germany alone, assigning scores based on many factors, including physical attributes and the quality and intensity of aroma. They found a strong linear relationship between the highest scores for aroma and later picking dates.

Those hops weren't as pretty, but they were the ones Boston Beer wanted. The company helped fund the study that affirmed what Grinnell and Koch had already smelled. They met with growers to explain why they wanted hops picked later. "You'd see heads starting to nod," Grinnell said. Farmers already knew that when they picked later, the plants grew better the following year.

"This is the brewer taking responsibility," Grinnell said, referring to more than supporting the study. "You have to keep in mind, the structure of hop growing in Germany is much different than in the U.S. You have to take responsibility for your variety and how the farmer treats it."

Walking through the hop research station in Tettnang only days before the 2011 harvest would begin, Franz Wöllhaf talked primarily about his areas of expertise—agronomics and plant protection. He works for the local district of the department of agriculture. Considering different harvest dates was new, he said, but the importance was clearer that day than it had been the one before. Several Tettnang farmers had met with a group from Boston Beer the previous evening. "They really care about aroma," Wöllhaf said.

"The thing we select for at the end of the day is oil," Grinnell said. "We have to drive that home at the farm level."

Until recently most of those farms were in Germany or England (English farmers ship baled hops to Germany to be pelletized using the same process as other hops for Boston Beer), because few Samuel Adams beers included American hops. "American hops improved faster than I realized," said Koch, who admits to being very sensitive to "catty" aroma and flavor. "We'd been so focused on classic hops I didn't notice. Breeders were more interested in upping the alpha acids or duplicating noble hop aroma. My feeling was to use the noble hops themselves. The price didn't matter to us."

Samuel Adams Latitude 48 IPA includes five hops grown around the 48th parallel: Hallertau Mittelfrüh, East Kent Golding, Simcoe, Ahtanum, and Zeus, the last three brimming with American character. In 2011 the brewery sold a "deconstructed" 12-pack with two bottles of the *Latitude 48* made with all five varieties, then two brewed only with Simcoe, two with Mittelfrüh, and so on, making it easy for drinkers to compare the distinctively different Simcoe or Mittelfrüh hops or the surprisingly similar Ahtanum and Golding. Coincidentally, the IPA hopped only with the classically German Mittelfrüh won a gold medal at the 2011 Great American Beer Festival in the English-Style India Pale Ale category.

"We are actively ready for American hops," Grinnell said a few months after the 2011 harvest, referring also to hops from the Southern Hemisphere, since he would be headed in that direction for selection a few months later. That includes those with the sometimes-exotic flavors many drinkers have embraced, but he emphasized bigger and bolder doesn't matter unless it ends up in a glass of beer.

"I keep coming back to practical realities. How can we be more consistent?" Grinnell said. "It's rarely the exotic that makes a difference."

Hop Quality Group: A Learning Process

Boston Beer is one of 12 members of the Hop Quality Group, an organization formed by brewers who in 2010 recognized a need for American craft breweries to communicate with American hop growers and to take responsibility for assuring the quality of their aroma hops. The tagline on the the HQG logo reads "oils over alpha," although any particular member is as likely to say "aroma over alpha" when talking about the short history of the group.

"We realized we didn't know everything we need to know ... how little we recognized what goes into producing quality hops," said Firestone Walker brewmaster Matt Brynildson. He has literally traveled the world on behalf of the Hop Growers of America and has been finding escalating interest in American hops with "special" aromas.

Discussions about forming the group began in earnest at the 2010 Master Brewers Association of the Americas annual convention. After Anheuser-Busch and InBev merged in 2008 the new company disbanded a field program established in 1987. That program, craft brewers quickly realized after it ended, was instrumental in maintaining the quality of aroma hops. About the same time Val Peacock, who worked as manager of hop technology at A-B from 1989 through 2008, retired. "Part of our conversations were when Val leaves … we were going to be left without a conduit for discussing quality," Brynildson said.

"Val's the smartest hop guy I know," said John Mallett of Bell's Brewery, who became the group's president. The first thing the original members did was to retain Peacock as a consultant.

The first goals the group members outlined included:

- Interface with hops growers and let them know what brewers such as themselves want from the hops they buy. "We don't care about alpha acids. That's what the communication was about in the past," Mallett said.
- Fund research focused on improving flavor and aroma qualities of hops. The first project, starting with the 2012 harvest, focused on drying practices and the effects of decreasing kilning temperatures on oil content and quality. (Research by Hop Products Australia indicates lowering drying temperatures in combination with delaying harvest dates benefits aroma hops. See p. 99.)
- Assemble a best practices document for growers. As with the kilning study, the group planned to make it available to the entire hop and brewing industry.
- Provide signs that emphasize hops are a food product for any farm that wants them.
- Visit farms and processing sites.
- Sponsor a "Cascade Cup" competition. The award goes to the farmer who grows the best Cascade hops each year. (Growers in several other countries, such as England and Germany, judge hops annually. The Hop Growers of Washington awarded a trophy to the Hop Man of the Year until the mid-1960s.)

"It's been an education for me," Peacock said. "I'd always looked at it through the eyes of a big brewer. The needs of the smaller brewers will be a little different." For instance, sanitation becomes more of an issue for brewers who dry hop extensively, because they are using hops closer to a "raw" form, without the benefits that result from boiling.

"Not only am I excited as a hop grower," said Oregon farmer Gayle Goschie, talking about Peacock and the new organization. "I can see how excited he is and (looking forward) to putting to use all these great ideas he has."

Pellets: Easier to Store but Just as Fragile

Big as they are, the hop rooms for the east and west breweries at Sierra Nevada Brewing in Chico, California, hold only hops that will be used almost immediately. Most of the hops the brewery keeps on site are stored deeper in the brewery, in a giant vault held at freezing temperatures. Bales of more fragile varieties, such as Crystal and Citra, have been divided into smaller portions and wrapped in Mylar to further protect them.

Average Alpha-Acid Loss One year stored at 68° F (20° C)	
Product	Loss
Whole hops	Up to 100%
Pellets	10%-20%
CO_2 extract	2%-4%

Pellets, the form of choice at most breweries, take up far less room, and when packaged properly may be stored for extended periods. They nonetheless require as much care. Because the pellet die crushes the lupulin glands, both bitter and aroma compounds in pellets oxidize three to five times faster than in cones, which is why processors package them in oxygen barrier bags, usually foils. These may be hard packs under vacuum or soft packs under nitrogen and/or CO_2.

Peacock provided brewers concise guidelines during a 2010 presentation at the Craft Brewers Conference:

- Unopened pellets retain most of the alpha for five years when frozen (26° F or -3° C), two to three years when refrigerated (40° F or 4.4° C). Aroma changes will occur more quickly, but at 26° F aroma remains relatively stable.

- Because unprotected pellets store more poorly than cones, opened foils must be used quickly. Cascade pellets should be good for two weeks at 50° F (10° C), and five weeks at 26° F (-3° C), longer if colder. Resealing the bag does not provide the same protection as the original package but will help retain aroma.

Peacock also warned brewers about deterioration that may occur during transit, citing a 2002 article in *Brauwelt International* by Adrian Forster. Some key points:

- Low volatile aroma components may increase during inert storage, resulting in a solvent smell, particularly at higher temperatures. The formation of gas may cause the package to balloon, enough that it will burst. Even when packages don't rupture, the contents may be damaged.
- Warm conditions over 77° to 86° F (25° to 30° C) should be avoided.
- The German study, of course, included concerns about overseas transportation, but the temperature within a truck standing in the sun (or sitting over a weekend) may also easily increase above 95° to 122° F (35° to 50° C). The research determined that temperatures 77° to 86° F (25° to 30° C) were acceptable if they lasted less than five days; 86° to 95° F (30° to 35° C) for less than two days; that 95° to 104° F (35° to 40° C) was dangerous; and above 104° F (40° C) unacceptable.[1]

Brewers may easily spot the problem if foils arrive ruptured or swollen. However, they usually deflate as they cool, so the damage may not be obvious. Examining the contents of each package when it is used remains essential. Spoiled pellets will usually appear brighter, like fresh green vegetables, and may smell solvent or herblike.

Polyphenols and Phenols

Polyphenols and phenolic compounds play a complex role in beer flavor and stability, further complicated because of both positive and negative contributions. Malt furnishes approximately 70 percent of beer polyphenols, although the hop contribution may increase with the addition of lower alpha hops. For instance, Hersbrucker introduces 11 times more polyphenols into wort with the addition of the same quantity (obviously more hop mass) of alpha than the higher alpha Magnum. The differences are both alpha and variety dependent.

Hop polyphenols enhance flavor stability because of their antioxidant qualities, which suppress the formation of undesirable staling compounds. As any brewer making heavily dry-hopped beers may testify, they also provoke beer haze. Beer drinkers in parts of the United States consider slightly cloudy beers they know have been dry hopped as acceptable. However, in other parts of the country and other countries, colloidal haze may be viewed as a fault, and brewers may choose to use polyvinylpolypyrrolidone (PVPP), but then lose some of the benefits of polyphenols.

The quantity of polyphenols that remains in much of the beer brewed worldwide has shrunk dramatically in the last 50 years because of the introduction of hop extracts and isomerized products, along with the increasing levels of alpha acids in hops and the decreasing levels of bitterness in beer.

Forster explained the impact in a paper presented at the First International Brewers Symposium on Hop Flavor and Aroma in Beer. When brewers reduced bitterness from the range of 20 to 25 IBU to 18 IBU, they cut the hop addition from 100 to 60 milligrams of alpha acids per liter, but, more dramatically, the levels of polyphenols from 50 to 4 milligrams per liter. "In many beers a hop polyphenolic character is lost, in most cases more or less unintentionally," he wrote.

He argued that this runs counter to trends in the food industry. "The high reputation of wine is also linked to 'healthy polyphenols.' It is a pity that the public image of beer is lower than that of wine," he wrote. "Apart from polyphenols, iso-alpha acids and isoxanthohumol also show promising health benefits. ... The trend towards lower bitterness in beer stands in contradiction to the potential health benefits of using more hops."[2]

This discussion is not necessarily relevant to brewers who use hops in a "whole" form, either cones or pellets, although it's worth considering when writing recipes that include hops or hop forms with lower levels of polyphenols—at least when their qualities are desired.

'Skunky' by Any Other Name ('Imported') Is Still a Fault

Although many beer drinkers associate "skunky" character with imported beers and consider it a positive characteristic, brewers recognize light struck flavor as a negative and that it clashes with intended aroma and flavor. German chemist Carl Lintner first described the odor and gave it the name "light struck flavor" in 1875, and in the 1960s Yoshiro Kuroiwa and associates in Japan determined 3-methyl-2-butene-1-thiol was the main source. MBT, in fact, closely resembles the odor that skunks spray and is derived from photodecomposition of isohumulones in the presence of a photosensitizer, riboflavin. It has a very low threshold of detection and occurs quickly in direct sunlight so may be perceived by the end of a pint consumed outdoors on a sunny afternoon.

Kuroiwa's work showed the blue part of the visible spectrum is most efficient in generating light struck flavor, which is why brown bottles provide protection from MBT, clear bottles don't, and green bottles only limited defense. Some breweries, for the most part large and often intent on packaging at least some of their brands in clear or green glass bottles, use advanced hop products to inhibit the formation of MBT. Recent research found two more "skunky" odor compounds that result from light-exposed iso-alpha acids, and that advanced hop products don't offer protection for these, making the term light-stable hop products not altogether accurate.[3]

Brewers who use hops in a traditional manner accept that when light and iso-alpha acids meet, "skunky" characteristics result. Kegs and cans are obviously excellent barriers to light, but brown bottles provide sufficient protection, particularly when they are stored away from light.

Some Like Their Hops Slightly Aged, Some Quite Old

Oxygenated compounds begin to form in the cone while hops still hang on the bine, contributing to desirable hop aromas, including those some call "noble." Bales once routinely sat at ambient temperatures on the farm for at least six to 12 weeks before they were transferred to cold storage or processed into another form. Many brewers continue to prefer such "moderately aged" hops. "It's a matter of religion," Peacock said. The quality of bitterness suffers with any aging, but some brewers want the different aroma compounds that result. He likened the process to curing tobacco, explaining that compounds form that do not exist in the hops themselves.

Classically, brewers in Belgium use hops aged even longer in *lambic* beers, hops that on their own smell cheesy or sweaty. Research in Japan confirmed the additional age not only reduces alpha acids but produces beers with additional citrus and fruity character. Outlining the results in his doctoral thesis, Toru Kishimoto suggested that esters are synthesized during the fermentation process from the substrate short-chain acids, which were generated by the oxidative degradations of humulone or lupulone.

The Japanese researchers brewed four beers, two with hops aged for 30 days at 104° F (40° C) and two with hops stored at 39° F (4° C). Two were dosed at the beginning of the boil, one with a fresh hop and one with an aged one. Two were hopped for aroma after the wort cooled, again with a fresh hop and an aged one.

The ones that included aged hops had increased concentrations of many fruity esters, particularly ethyl 2-methylbutanoate (citrus-apple), ethyl 3-methylbutanoate (citrus-sweet-apple), and 4-(4-hydroxyphenyl)-2-butanone (citrus-raspberry). Additionally, the concentrations of green, hop pelletlike, and resinous components, such as myrcene, were lower. Kishimoto found that overall the balance of compounds changed in a way that accented the citrus character of the beers made with aged hops.[4]

Although oxidation generally does little good for hop aroma, brewers interested in particular esters, and certainly those brewing traditional *lambic* beers, may benefit from controlled oxidation.

Dry Hopping and Flavor Stability

Hops provide and protect positive beer aromas and flavors until they don't. Dry hopping and other post-fermentation additions confuse the equation. When brewers dry hopped in casks and publicans served those casks in a matter of days, they did not need to be concerned about shipping or long-term flavor stability.

"(Pleasant hop aroma) is like the legs of an athlete, the first thing to go," said Dan Carey of New Glarus Brewing. He senses one of two things happening when heavily dry-hopped beers grow older. "They either begin to produce flavors I don't like, or the flavor is lost."

In the second situation, the resulting beer may be perfectly pleasant although it lacks some of its original character. In the first, the beer is ruined. "Oxidized Cascade is not something I enjoy," Carey said.

Both a study funded by Boston Beer and the Barth-Haas Group that examined the influence of harvest dates on flavor stability in dry-hopped beers, and similar research in Belgium that included in-depth analysis of late hop additions, focused on lager beers and the role of hop aroma in masking stale flavors attributed to oxidative reactions.

The research in Germany evaluated Hallertauer Mittelfrüh hops grown on four farms in the Hallertau region and picked on five different dates, spanning more than three weeks. Beers brewed with hops from the last date had slightly better flavor stability, but with increasing storage time and temperature the difference became minimal after 45 days. Just as aroma and flavor impact varied depending on where the hops were grown, so did stability. Bitterness intensity did not substantially decrease with increased storage time or temperature, and the authors of the study suggested the influence of hop aroma on perceived bitterness as a possible explanation.[5]

The Belgian study found that late hop additions, which included both conventional (pellets) and advanced hop products, lowered the perception of staling aldehydes that analysis indicated were present. The authors wrote, "This finding points to the potential of hop aromatization to mask aged staling flavors. Indeed, as mentioned earlier, the conventionally late-hopped brew scored high for hoppy aroma in the fresh beer."[6]

More recent research in Belgium revealed that flavor stability may be variety dependent. Thus, some varieties are better used for dry hopping, others are best added late in the boil. A team from the Proef Brouwerij and the Laboratory of Enzyme, Fermentation, and Brewing Technology at KaHo St.-Lieven/Ghent evaluated six single-hop beers brewed at Proef. They used three varieties of hops, comparing late-hopped and dry-hopped versions of both. They analyzed and evaluated beers after one month, three months, and six months.

The level of myrcene was much higher in the dry-hopped beers and decreased with aging, while it remained fairly constant in the late-hopped beers. Humulene and caryophyllene behaved much the same. The sum of humulene epoxides was slightly higher in the dry-hopped beers and did not change significantly during aging. As is typical, the level of linalool increased.

After six months, Strecker aldehydes (which may be perceived as fruity, sweet, and honeylike) and furfural (almonds) levels associated with aged beers increased much more in two dry-hopped beers, called A and B, but were lower in beer C. Dissolved oxygen (DO) and headspace oxygen (HSO) were found to be lower in late-hopped beer A and B, while the DO in C was comparable and the HSO lower in the dry-hopped beer. Trained panelists noticed little difference in "aging effects" in fresh beers or those a month old. Distinct differences emerged by six months. Panelists gave dry-hopped beers A and B higher (therefore worse) aging scores than late-hopped A and B, but late-hopped C a higher score than dry-hopped C. Not surprisingly, the panelists indicated a personal preference for the beers with lower aging scores.

The researchers concluded, "The impact of an additional dry-hopping step on single-hop beer flavor stability is cultivar dependent. The interplay of the degradation of iso-alpha acids and hop oil-derived constituents on the one hand, and the formation of staling aldehydes on the other hand, is obviously a very intricate issue. Whether oxygen content plays a significant role in the observed analytical and sensory changes upon aging is unclear, but in any case, the dry-hopped beer C shows a markedly enhanced sensorial flavor stability compared to the other dry-hopped beers. Further research is needed in order to confirm these observations and to elucidate the possible underlying (bio-)chemical reasons for these interesting findings."[7]

Yvan Borremans from Proef explained via email that they decided not to disclose the varieties, because "in time we want to come forward and publish these results in detail, but also because the results are based on one single-hop brewing campaign so far. Therefore, we prefer to err on the cautious side, and we would like to confirm our observations before we share more detailed information.

"Flavour stability is in fact a combination of the decline of wanted flavours and the increase of unwanted flavours over time. All sorts of synergistic and antagonistic (masking) effects can take place between each pair of flavour-active compounds involved. Needless to say, any assessment of these effects is extremely complex given the sheer multitude of compounds found in beer. DO is surely relevant for the oxidative degradation of iso-alpha acids and lipid oxidation, and, as such, it can contribute to the formation of staling aromas. But does DO explain the whole picture that we are seeing? I would say it does not; just look at furfural. We obviously have thoughts and ideas in this respect, but it would be premature to elaborate at this time."

Packaging, which is for the most part beyond the scope of this book, may also change aroma impact. Sierra Nevada Brewing discovered its filter pads scalped a significant amount of odor compounds from its beers and worked with companies to change the pads. Crown liners also scalp aroma. "The first thing I do when I open a beer is smell the liner," said Tom Nielsen of Sierra Nevada.

Levels of significant compounds that produce floral, spicy, and woody aromas drop dramatically the first three days after bottling, as they migrate from the liquid to the head space to the liner, and perhaps eventually into the atmosphere. After three days an "average" IPA might contain the same level of myrcene, for instance, as a pale ale immediately after bottling. How fast the aroma continues to fade in the following weeks depends on many factors, including storage temperature and how much the beer is agitated in shipping.

"Every (liner) scalps," Nielsen said. "You have to decide which (aromas) you want to give up." Those who have trouble controlling the level of dissolved oxygen in packaging may pick a liner that actively scavenges oxygen, but that also means losing more desirable hop aroma.

The compound 4-mercapto-4-methylpentan-2-one (4MMP), which is a major contributor to "American" hop aroma, is particularly volatile. It may also be perceived as catty (an aroma found in the early stages of oxidation) by some drinkers but more likely will retain its pleasant aromas when consumed fresh.

Firestone Walker brewmaster Matt Brynildson uses Chinook as an example of how oxidation changes the aroma of raw hops. "It's really grapefruity, pungent, when it's fresh. I love it," he said.

"Open a bag, put in the fridge," he said. Even in a cold environment it takes only a short time for that aroma to turn into one that smells intensely "catty," just like a dirty litter box.

"It's obvious when you know the beers," Brynildson said, discussing what happens to hop flavor and aroma as beer ages. "It's oxidized, but how do you describe that (aroma)?"

In the best case, bright, hoppy aromas simply fade, or "lose their legs." "The hoppier the beer, the shorter the shelf life. At 30 days I'll notice the difference," Brynildson said about his own beers, which he obviously knows particularly well. "Most of these (beers) need to be handled cold all the way through."

Notes

1. Adrian Forster, "What Happens to Hop Pellets During Unexpected Warm Phases?" *Brauwelt International* 2003/1, 43-46.

2. Thomas Shellhammer, ed., *Hop Flavor and Aroma: Proceedings of the 1st International Brewers Symposium* (St. Paul, Minn.: Master Brewers Association of the Americas and American Society of Brewing Chemists, 2009), 123.

3. Christina Schönberger and T. Kostelecky, "125th Anniversary Review: The Role of Hops in Brewing," *Journal of the Institute of Brewing* 117, no. 3 (2011), 265.

4. Toru Kishimoto, "Hop-Derived Odorants Contributing to the Aroma Characteristics of Beer," doctoral dissertation, Kyoto University, 2008, 72-75.

5. G. Drexler, B. Bailey, C. Schönberger, A. Gahr, R. Newman, M. Pöschl, and E. Geiger, "The Influence of Harvest Date on Dry-hopped Beers," *Master Brewers Association of America Technical Quarterly* 47, no. 1 (2010), doi:10.1094 /TQ-47-1-0219-01, 4.

6. F. Van Opstaele, G. De Rouck, J. De Clippeleer, G. Aerts, and L. Cooman, "Analytical and Sensory Assessment of Hoppy Aroma and Bitterness of Conventionally Hopped and Advance Hopped Pilsner Beers," *Institute of Brewing & Distilling* 116, no. 4 (2010), 456.

7. Y. Borremans, F. Van Opstaele, A. Van Holle, J. Van Nieuwenhove, B. Jaskula-Goiris, J. De Clippeleer, D. Naudts, D. De Keukeleire, L. De Cooman, and G. Aerts, "Analytical and Sensory Assessment of the Flavour Stability Impact of Dry-hopping in Single-hop Beers," poster presented at Tenth Trends in Brewing, Ghent, Belgium, 2012.

10

What Works

Theory aside, what matters is what ends up in the glass

Hans-Peter Drexler's German brewing education focused (not surprisingly in the 1970s) on bottom-fermented beers and styles such as hop-accented German *Pils*. He learned "Bavarian wheat beer brewers were known as the strange people who carry the hop bag alongside the wort kettle and do not put it in." Shortly thereafter he went to work at Private Weissbierbrauerei G. Schneider & Sohn, the largest brewery anywhere that is still making wheat beers using strictly traditional methods.

Schneider Weisse Original balances a glass full of fruity esters and spicy phenols with 14 bitterness units from hops grown in the nearby Hallertau region. Getting the hops right doesn't have to be difficult, but it is important. It is one thing to consider the compounds a single hop produces, another to blend several together with malt, then to add yeast. To explore the role of hops in various styles extensively would take another volume and is a reason those styles merit their own books. Instead, the recipes that follow illustrate how a few brewers include hops within the context of what we really care about—beer.

Given a chance to brew with two experimental German varieties early in 2012, Bear Republic Brewing brewmaster Richard Norgrove started with a base best described as a wheat wine. Norgrove added considerably more hops, and his beer finished with between 70 and 75

IBUs (calculated). He blended Mandarina Bavaria and Polaris in a ratio of 60 to 40 or 40 to 60, depending on the addition, making one at the beginning of a 90-minute boil, one with 60 minutes remaining, one with 40, and then dry hopping with the pair.

"I like to do a lot of blending, maybe change the way the oils come across," he said. He talked in terms of abstract art versus portrait art, probably because he paints with watercolors himself. "With watercolors you dilute or strengthen the vibrancy of color by the way you use water."

Even the recipe for a well established beer, like the immensely popular *Racer 5 India Pale Ale,* needs adjusting on a regular basis because of variations from one hop-growing year to the next and one region to the next. "I grow Centennial in my yard. In the back of the house and in the front of the house it is entirely different," he said. "So how do I make *Racer* taste the same when hops change? That's what makes it craft."

He hauled out a recipe for a 17-barrel batch from 2005 to illustrate, the hop addition made during and after a 90-minute boil:

90 minutes: 2.5 pounds each of Chinook and Cascade

60 minutes: 2.5 pounds each of Cascade and Centennial

Whirlpool: 5 pounds of Cascade and one pound of Amarillo

Dry hop in primary: 4 pounds of Cascade and 4 pounds of CTZ

Hop baby (later added to finished tank): 1 pound each of Cascade, Centennial, Amarillo, and CTZ. The "hop babies" are basically hop tea vessels, and Bear Republic uses eight 30-barrel babies for dry hopping in its production brewery, each serving a 300-barrel tank.

"There are multiple ways to change the perception in your mouth," Norgrove said. "My goal is to capture it for the whole beer, from start to finish. How do I make this excite me from beginning to end?"

Racer 5 contains as many hop varieties as Russell Schehrer used the entire first year he made beer at Wynkoop Brewing in Denver. Schehrer was a groundbreaking brewer—the Brewers Association named its annual award for innovation in his honor—but less than 20 years ago pub brewers had a limited number of choices. In the 12 months after he and John Hickenlooper started Wynkoop in 1988 Schehrer used only Cascade, Hallertau (probably Mittelfrüh), Tettnang (likely U.S.), Willamette, and Bullion.

To brew an India pale ale he added Bullion at the outset of a 90-minute boil, more Bullion with 30 minutes left, and Willamette and Cascade five minutes before the boil ended. He dry hopped it in the serving tank with a blend of Cascade pellets and cones plus French oak chips.

Wynkoop brewmaster Andy Brown entered 2012 with a list of a dozen hop varieties he planned to use regularly and several others he intended to experiment with. He looked at names new to him and admitted he wasn't quite sure what to expect.

"You actually have to brew with hops to figure them out," said Vinnie Cilurzo at Russian River Brewing. He started using a recipe for that purpose when Russian River was still located in Guerneville, California, naming the beer *Hop 2 It*.

"The idea behind the beer was to have the exact same malt bill and hop bill, where the only thing changing is the actual hop variety and the quantity of the first hop addition, which was changed only to match the bitterness from batch to batch. I started out brewing with lots of old school varieties, Eroica, Bullion, Bramling Cross, Brewer's Gold, etc.," he said. "Then I moved into the more common hops, Cascade, Centennial, Chinook, CTZ, etc. From there I went to the new varieties and was lucky enough to brew with several hops that were in their experimental stage that are now in the trade. This included what became Palisade and Simcoe."

Simcoe, of course, became an essential part of Russian River's *Pliny the Elder*, and in turn *Pliny the Elder* helped make Simcoe a popular variety. He still uses the recipe when he tests experimental varieties at the request of breeders and farmers or considers using a new one in his brewery.

"Brewing with single hops rarely yields a beer that could be an actual ongoing recipe, but it does teach you what component of the hop works and does not work," he said. "For example, I remember brewing with Amarillo in *Hop 2 It*; it had an awful bitterness, but the flavor and aroma were just fantastic. I've never used it for bitterness since."

Hop 2 It
R&D single-hop brew

Original gravity: 1.052-1.056 (12.9-13.8 °P)
Final gravity: 1.010-1.012 (2.6-3.1 °P)
IBU: 30-40
ABV: 5.5%-5.8%

Grain bill:
74% 2-row malt (domestic)
13% Maris Otter malt (English)
10% crystal 20 malt (domestic)
3% acidulated malt (German)

Mashing:
Single-infusion mash at 154° F (68° C)

Hops:
90 minutes (5 to 10 IBU)
30 minutes (20 IBU)
0 minutes (10 IBU)
Dry hop, 1 week at 68° F (20° C), variable

Boiling: 90 minutes
Yeast: California Ale
Fermentation: 68° F (20° C)
Packaging: Target 2.5 volumes CO_2 (5 g/L)

Instructions: Use the same malt bill each time. Keep the dry hop, final hop addition, and middle hop addition the same quantity each time. The only variable between each single-hop batch is the hop variety itself and the quantity of the first hop addition, which will be based on your calculations to hit the target IBUs. So, a lower alpha acid hop will have a higher 90-minute hop addition, and vice versa for a higher alpha hop.

About the Recipes

Legal disclaimer: The recipes that follow are offered "for educational purposes only." I've tried to present them in a way that makes it easy for a brewer to visualize how they might work at her or his own brewery, but also so it is clear how they are used within the original brewery or to reflect the contributor's intent. That's why, for instance, gravities are often listed first in Plato, because that is the measure most commercial breweries use, and conversions are not always exact.

Fact is, although the first hop addition in the recipe from Meantime will add about 12 IBUs the recipe calls for five kilograms because hops come in that size package. The quantity converts conveniently to 1.5 ounces per barrel purely by chance, and works best for a brewery that, like Meantime, expects 35 percent utilization on a 75-minute addition.

Recognizing such variations, Ron Barchet of Victory Brewing simply specified bitterness targets for each addition, leaving it to a brewer to calculate the proper amounts, because "it would be impractical to attempt to give you actual amounts of hops, as everyone's brewery yields different results, dependent on water chemistry, wort pH, boiling methods, and hop analytics."

As those who use the metric system will point out, conversions are more straightforward when they don't include mixing barrels, gallons and ounces. Key ones to remember are that 100 grams per hectoliter (1 g/L) equals about 85 grams per barrel, which is comparable to three ounces. One pound per barrel is equivalent to 3.85 grams per liter.

Recipes change all the time at commercial breweries. John Keeling said that Fuller's in London will make adjustments during the course of a brewing year based on input from its tasting panel. "All the numbers can be right, but maybe not the taste," he said. "The panel will notice bitterness coming through a bit more. We'll tone it down."

Sierra Nevada Brewing does not always use the same hop variety for the bittering addition in its pale ale, which is about as iconic as an American craft beer can be. "I've got contracts for multiple varieties, so I can pick and choose across quality," brewmaster Steve Dresler explained. "I have a Perle contract, one for Sterling. We work with the dealers, so we are using the best quality hops available."

The Recipes

Brasserie de la Senne
Brussels, Belgium

Yvan de Baets eschews the words "beer style," but ask him the question most brewers duck, what his favorite is, and he's quick to answer. He loves bracing, bitter beers from Great Britain, not exactly surprising given how much he appreciates bitterness itself (p. 182), a quality apparent in Brasserie de la Senne beers.

"This is art, this is what I like to do," he said, raising a glass to his lips at the Great British Beer Festival in London.

He enjoys many beers brewed with new varieties but called this recipe "Old World's Mantra" for good reason. "In all of Europe it seems the trend is to use New World hops. When everybody does the same thing, it become a bit boring," he said. "Maybe special becomes less special."

He appreciates the quality British brewing water contributes but loves the hops on their own as well. "They have such finer and subtler aroma," he said.

He's not afraid to used Old World words such as "balance" and "drinkability." "If this recipe gives a hoppy beer it is in balance," he said. "To me a beer should replace water to refresh you. A little bit mood altering, too, but I don't want to go too far."

He smiled, probably thinking bitter thoughts.

Old World's Mantra
Original gravity: 12 °P (1.048)
Final gravity: 3 °P (1.012)
IBU: ~50
ABV: 5.2%

Grain bill:
82% Pilsner malt
12% Munich malt (25 EBC)
6% crystal malt (120 EBC)

Mashing:
50° C (122° F), 15 minutes
63° C (145° F), 45 minutes
72° C (162° F), 20 minutes
78° C (172° F), 5 minutes

Hops:
Challenger, beginning of boil, 300 g/hL
Styrian Golding, 10 minutes before end of boil, 150 g/hL
Styrian Golding and Bramling Cross, whirlpool or hop back, 100 g/hL
each

Water: Hard
Boiling: 75 minutes, hard boil
Yeast: Neutral. Highly attenuative. Flocculent
Fermentation: Main: Maximum temperature 26° C (79° F). Lagering: 4
weeks at 10° C (50° F)
Packaging: Refermentation in the bottle. 2.5 volumes CO_2 (5g/L)

Dogfish Head Craft Brewery
Milton, Delaware

Founder Sam Calagione calls this recipe the homebrewer-friendly version
of *Indian Brown Ale*, with pretty much the same DNA as Dogfish Head
Craft Brewery's commercial version.

When Calagione started bottling beer in 1996 he packaged *Shelter Pale
Ale, Chicory Stout,* and *Immort Ale.* Customers asked for a hop-forward
beer. "I wanted to find an off-centered way to do this," Calagione said.
"I decided to use hop varieties, addition times, and volumes that would
complement the bitter notes of dark grains."

He called it *Indian Brown Ale* to signal it had the hop character of an
India pale ale and "as a subtle nod to the flaked maize we use to keep the
body from being too fat," he said.

"If the term 'black IPA' existed when we came out with this beer in
1997 I guess that is the category it would have fit into. So it's sorta like a
prototype black IPA," he said, laughing hoarsely.

Indian Brown Ale
Original gravity: 1.074 (18 °P)
Final gravity: 1.020 (5.1 °P)
IBU: 50
ABV: 7%

Grain bill:
71.3% 2-row malt
6.4% amber malt
6.9% crystal 120 malt
7% flaked maize
3.5% coffee
1% roasted barley
3.9% brown sugar

Hops:
Warrior, 60 minutes (33 IBU)
Simcoe, heat off or whirlpool (17 IBU)

Mashing:
Treat your water as you would for a Burton-style IPA. Adjust your water temperature to hit 156° F (69° C). Add the 2-row first to build a base, then the remainder of the grains. With the use of maize, a 113° F (45° C) protein rest may be desired if you have the capability to raise to conversion temperature. Hold for 30 minutes. Sparge at 169° F (76° C) with a slow runoff rate due to the maize and specialty malts.

Boiling: 60 minutes
Yeast: California Ale or another low-ester-producing yeast
Fermentation: Cool to 60° F (15.5° C) and hold cooler if possible during ferment at 66° F (19° C). Rack to secondary and dry hop when fermentation is complete, and hold for 21 days.

Dry hops: 0.5 pound/barrel of Vanguard for 21 days
Packaging: 2.6-2.7 volumes CO_2 (5.2-5.4 g/L).

Epic Brewing and Good George Brewing
New Zealand

New Zealand has no native hops. The hops grown there are a result of crossing imports from North America, England, and the European continent. They grow closer to the equator than in the largest Northern Hemisphere hop growing regions—at about 41-42° south, in a temperate coastal climate, with some farms only a few kilometers from the ocean. "The appellation is huge. That's why nothing grown in New Zealand tastes like anywhere else," said Dave Logsdon, who uses several New Zealand varieties at his brewery, Logsdon Farmhouse Ales, located a bit north of the Willamette Valley in Oregon.

Luke Nicholas (Epic Brewing) and Kelly Ryan (Good George Brewing) have a better idea than most of how New Zealand's brewers are using their hops, because early in 2011 they spent 17 days traveling 4,500 kilometers around the country. Brewers in other countries may call a beer brewed with Kiwi hops a "New Zealand Pale Ale" without realizing in its home country it has become a distinctive style.

"My interpretation is that they aren't super-malty. They have a good little caramel kick, a touch of color and richness to balance out the big, grassy, mouth-coating bitterness that NZ hops can give," Ryan said. "I think a small amount of residual sweetness helps with this balance, especially as some NZ hops can give quite a lingering after-bitterness. NZ pale ales are all about the fruit. Passion fruit, gooseberry, lychees, mangoes, and that lovely 'greenness' that makes you close your eyes and think of our forests and fields."

New Zealand is not among the largest dozen hop growing regions, but mostly because of demand for hops with special aromas, New Zealand Hops established a five-year program in 2012 that would boost production by almost 30 percent. In addition to adding acres, farmers have begun to replace higher alpha hops with special varieties such as Nelson Sauvin, Motueka, and Wakatu.

Some varieties will remain difficult to find in the shorter term. For instance, NZ Hops basically removed Riwaka from the export list. Spokesman Doug Donelan explained in an email: "Riwaka is a high demand variety with limited volume currently in production. We need to rationalize distribution to ensure current users aren't disadvantaged while trying to expand acreage. The U.S. is only a small market for Riwaka, with only one brewer currently using any significant volume. Existing users will continue to be supplied. We are just limiting our expansion for the time being."

New Zealand Pale Ale

Original gravity: 1.055 (13.6 °P)
Final gravity: 1.012 (3 °P)
IBU: 40
ABV: 5.6%

Grain bill:
In percentages
75% pale malt (Gladfield or Maris Otter)
10% Munich malt (1-15 EBC)
8% Caramalt
2% pale crystal malt

Mashing:
153-154° F (67-68° C) single infusion

Hops:
Pacific Jade, 60 minutes (7-8 IBU)
Pacific Jade, 30 minutes (7 IBU)
Motueka, 30 minutes (3 IBU)
Wait-ti, whirlpool (11 IBU)
Motueka, whirlpool (7 IBU)
Pacific Jade, whirlpool (4 IBU)
(Pacific Jade is a 12-14% alpha hop with Saaz and Cluster as ancestors. Wai-ti is a new, low alpha/high oil variety in short supply.)

Boiling: 60 minutes
Yeast: Wyeast 1272
Fermentation: Pitch at 17° C (63° F), ferment at 19° C (66° F), making sure it does not rise above 21° C (70° F). Fermentation should take 4-7 days. Reduce to 4° C (39° F) during days 5-6. Hold, make second dry-hop addition, do not dry hop for more than 7 days.

Dry hops: Blend Wait-ti, Motueka, and Pacific Jade in equal parts, 5 g/L. Add half when beer is near target gravity. Add half after racking beer off yeast into secondary.

Packaging: 2.5 volumes CO_2 (5 g/L)

Firestone Walker Brewing
Paso Robles, California

When brewmaster Matt Brynildson began working on the recipe for an India pale ale in 2006 he envisioned it might be brewed with English malts and would be fermented, like many other Firestone Walker beers, using the brewery's unique Union system and spending time in oak barrels.

He was wrong.

The brewers at Firestone Walker made test batches for the better part of a year. "The first brews were maltier, sweeter, not what we were going for," Brynildson said. The first thing to go was the British malts. "We weren't trying to hold ourselves to a (specific version). We were going to make the best possible IPA."

Firestone's Union takes inspiration from Britain's classic Burton Unions, the last of which is located at Marston's in Burton-upon-Trent. The brewery uses new American oak barrels, and rotates fresh ones in regularly, so they add considerable oak character.

The base for what will become Firestone Walker *Pale 31* and FW *Double Barrel* ferments in steel. Approximately (depending on several factors) 20 percent is transferred into the Union after the first day of fermentation, and that remains in wood for seven days. What's blended back with beer fermented in steel becomes *Double Barrel Ale*. The brewers then blend about 15 percent of *Double Barrel* with unoaked pale ale to create *Pale 31*.

"We were trying to integrate the oak (in the IPA), but we just gave up," Brynildson said.

Instead they created a beer that quickly became a benchmark for "West Coast American IPA." *Union Jack* won a silver medal at the 2008 World Beer Cup shortly after it was released, then later in the year a gold medal (which it also won the next year) at the Great American Beer Festival and a gold medal at the European Beer Star awards.

Union Jack IPA
Original gravity: 16.5 °P (1.068)
Final gravity: 3.0 °P (1.012)
ABV: 7.5%
IBU: 75
The discussion of ingredients and process is best understood in Brynild-son's own words:

Grain bill:
88% American or Canadian highly modified 2-row malt (we utilize Rahr)
6% Munich malt (I use a similar blend in all of our pale ales to replicate English pale ale malt up to 15%)
3% Briess CaraPils malt
3% Simpson 30/40 malt

Water treatment:
Add gypsum as needed to get total calcium above 100 parts per million (we use a little CaCl to increase finish Ca). I'm not big on overly Bur-tonized water. We start with RO treated water. Acidify mash to 5.4 with phosphoric or lactic.

Mashing:
Mash at low temp 145° F (63° C) for 45 to 60 minutes and step up to 155° F (68° C) to finish conversion. You might add dextrose (up to 5%) to get to gravity and/or to aid in attenuation if needed. The idea is to have a lean body, not to hide hop character, and to help accentuate the hop profile.

Hops, yeast, fermentation, dry hopping:
We whack the boil hard with Magnum, but I'm not overly anal about the bittering charge when it comes to variety when brewing IPAs and double IPAs. The charge we are using calculates out to be 50 IBU at 15% alpha. I've been known to use some hop extract (purified Isolone) in the kettle if we are having a hard time hitting the 75 IBU mark. It shouldn't be necessary, though.

I like middle additions on IPAs. We use Cascade at 30 minutes. Ama-rillo would work well here as well. The charge is 14 IBUs at 6% alpha.

We hit it again at 15 minutes before knockout with Centennial with the same amount as the Cascade addition. You might notice that we are already over the 75 IBU mark on paper, but utilization is low. It is all about hop flavor at this point.

We hit it in the whirlpool with another charge of Cascade and Centennial (equal amounts). On paper this looks like a 40 IBU charge but, again, the utilization of alpha on a brew like this is low overall.

We ferment UJ with our house ale yeast cooling in at 63° F (17° C) and setting the fermentation at 66° F (19° C). Our house ale strain is closest to London ale or other English-style (fruity/soft) yeast. When the brew reaches 6 °P (1.024) we turn the fermenter up to 70° F (21° C) for VDK (diacetyl) reduction and for the dry hop additions.

We dry hop the beer two times at about one pound per barrel. Once at 0.5-1 °P (0.002 to 0.004) before the end of fermentation (about day 5) and again three days later while the beer is still warm (prior to crash cooling the beer). We use a blend of Centennial and Cascade, with lesser amounts of Simcoe and Amarillo for each dry hop. I'm a firm believer in short contact time with the hops, no more than 3 days. The yeast and hops at the bottom of the fermenter are removed prior to each dry hop.
Packaging: 2.55 volumes (5.1 g/L)

Fuller, Smith & Turner
London, England

Fuller's first brewed *1845* at its Griffin brewery in Chiswick to mark the 150th anniversary of the Fuller, Smith & Turner partnership. More recently, the brewery introduced a Past Masters series of beers, made using recipes from the nineteenth century. "We have the records (recipe books that go back to 1845) and the ability to brew it at the same place it was brewed before," brewing director John Keeling said, standing beside a mash tun decommissioned long ago. "Though not the same equipment."

The brewers made small adjustments after the first batch. "There's a learning curve," Keeling said. "We toned down the hops a bit. Modern Goldings are more bitter, higher alpha." Much of the brewing year they also are fresher than in the nineteenth century.

Fuller's replaced flowers with pellets in 1976 and, naturally, discarded its hop backs. Keeling gestured again to the hand-operated mash tun, then away. "Everything on this side was knocked down and rebuilt," he said. Fuller's invested another £40 million for renovations in 1999. Production tripled between 1981 and 2011.

When Keeling went to work at Fuller's in 1974, the brewery fermented beer in both open tanks and closed conicals. Keeling asked the brewer who had worked at Fuller's the longest which made better beer. "He said, 'John, in actual fact (open fermenters) make the best beer, and they make the worst.' On the whole they weren't any better," Keeling said.

"Consistency comes from plant process. Better beer is based on ingredients, philosophy, and standards of quality. All people items," Keeling said.

Fuller's 1845
Original gravity: 1064.5 (16 °P)
Final gravity: 1015.5 (4.1 °P)
IBU: ~ 52
ABV: ~ 6.3

Grain bill:
78% Best pale ale malt
19% amber malt
2% crystal malt
1% chocolate malt

Water treatment:
Key parameters: reducing naturally high carbonate levels to less than 80 parts per million; sulphate additions to increase level above 200 ppm; calcium greater than 200 ppm

Infusion mashing:
Mash and sparge liquors at rates to produce wort capable of producing beer at sales gravity. This brew is not parti-gyled.

Hops:

Golding, start of boil. Approximately one pound per barrel to achieve final beer IBU.

Golding, late copper with short boil to mix, one-quarter pound per barrel

Boiling: 60 minutes, vigorous, with evaporation rate a minimum of 7 percent

Yeast: Fuller's own ale strain

Pitching rate: Approximately 4 pounds per barrel, to achieve more than 15 million cells/mL

Fermentation: Pitch at 62° F (17° C), allow to rise to 68° F (20° C).

Packaging: Bottle conditioned. 2.3 volumes CO_2 at start, 2.6 after conditioning (4.6 g/L and 5.2 g/L)

Kissmeyer Beer & Brewing
Denmark

Anders Kissmeyer spent 16 years working for international brewing giant Carlsberg. He started the Danish craft brewery Nørrebro Bryghus in 2001. Since he left Nørrebro in 2010 he has traveled much of the world, brewing Kissmeyer Beer at various locations and collaborating with friends in many countries on special beers. He also writes, teaches, and consults.

When he decides, "It's very much in my DNA as a brewer to take classical styles and try to give them a unique and personal twist," he's got a plan.

"I love a good imperial IPA, but I have to say that I am often disappointed when drinking this style," he wrote in an email. "To me, too many of these beers are cloying, sweet, and over malty. Not to mention the many that have an astringent and aggressive bitterness that cuts through your throat like a chain saw."

Stockholm Syndrom, which he first brewed at Sigtuna Brygghus outside of Stockholm, resulted from his fascination with the fruitiness of imperial IPAs. "It should be like drinking a hoppy, smooth, and crisp bowl of citrus and exotic fruits," he wrote, describing an explosion of citrus, peach, pineapple, and passion fruit.

Choosing the fruitiest hops he knew of was only the first step. He designed the recipe to showcase them by: a) making the malt grist flavor neutral, b) using sugar for greater attenuation, c) mashing for maximum fermentability, and d) picking a yeast with a neutral flavor profile that, again, would maximize attenuation. "In essence, the crispness is a combination of a smooth and pleasant bittering hop and then making the beer as dry as at all possible," he wrote.

Kissmeyer Stockholm Syndrom Imperial IPA
Original gravity: 20 °P (1.083)
Final gravity: 2.5 °P (1.010)
IBU: 100
ABV: 9.5

Grain bill:
55% lager malt (2-row base malt)
25% pale ale malt (preferably English)
13% CaraPils malt or similar (may be replaced by lager malt if not available)
3% pale wheat malt
3% pale crystal malt (~ 125 EBC)
1% dark crystal malt (~ 350 EBC)
10% of total extract from a neutral, 100% fermentable sugar (dextrose or similar)

Mashing:
a) Step infusion
Mash-in 50° C (122° F), 20 minutes, pH of mash: 5.3 to 5.5, adjust with acid if necessary
Heat to 64° C (147° F) in 15 minutes, 1° C per minute
Saccharification 64° C (147° F), 45 minutes
Heat to 70° C (158° F) in 15 minutes, 1° C per minute
Saccharification 70° C (158° F), 15 minutes, hold at 70° C (158° F) until iodine test is negative
Heat to 78° C (172° F) in 10 minutes, 1° C per minute
Mash-off 78° C (172° F), 15 minutes
b) Straight infusion
Mash-in 68° C (154° F), pH of mash: 5.2 to 5.4
Saccharification 65° C (149° F), 60 minutes

Hops:
Columbus, mash (10 IBU)
Green Bullet, mash (10 IBU)
Columbus, 60 minutes (30 IBU)
Green Bullet, 60 minutes (30 IBU)
30 grams per hectoliter of each:
Simcoe, 50 minutes
Green Bullet, 40 minutes
Pacific Gem, 30 minutes
Amarillo, 20 minutes
Simcoe, whirlpool
Pacific Gem, whirlpool
Amarillo, whirlpool
Nelson Sauvin, whirlpool
(*Pacific Gem is a 13-15% alpha hop bred from New Zealand stock, Cluster, and Fuggle.*)

Boil: 60 minutes
Yeast: American Ale, 15 million cells/mL
Fermentation: 72° F (22° C)

Dry hops:
Begin as soon after cropping yeast as possible, 14-16° C (57-61° F).
Length based on preference
Add 50 grams per hectoliter:
Simcoe
Nelson Sauvin
Pacific Gem
Vanguard

Secondary: After dry hopping, beer is cooled to 8° C (46° F) for 5-7 days. The temperature is reduced to 0° C (preferably as low as -1.5° C; 29-32° F). The cooling rate is not critical, but the beer must be lagered at the low temperature until clarity and flavor is satisfactory.
Packaging: Do not filter. Target 2.5 volumes (5 g/L).

Marble Brewery
Albuquerque, New Mexico

The good news for Marble Brewery when it opened in 2008 was *Marble India Pale Ale* became an instant hit, and the Albuquerque brewery sold nearly 5,000 barrels of beer in its first full calendar year of operation. The bad news was that hops became much more expensive in 2008 than they were in 2007, and contracts for future deliveries became more important.

Director of brewing operations Ted Rice quickly discovered there was more to getting the hops he wanted than writing contracts. "I learned they aren't all the same quality," he said. He is more sensitive than most to sulfur compounds that create garlic, onion, and petroleum odors. "I don't want to write a big contract and discover I don't like that hop."

He recognizes that even the most stable varieties won't be the same every year. "Beer's an agricultural product. There will be a shift, but you don't want to get into this wine way of thinking—'That's a good batch'—like it's out of your control." He certainly wouldn't hesitate to make hopping changes to keep his beers consistent.

As the recipe here illustrates, staying consistent doesn't mean avoiding new varieties. "You learn by staying in touch with the industry. Rubbing hops, drinking other people's beer," he said.

Marble Red Ale drinks in particular contrast to *Marble India Pale* (p. 29), which has a lean body that stays out of the way of lush, fruity hops. Rice said he wrote the recipe for the *Red* while drinking a beer in the Marble tasting room and thinking about sweet flavors that would support a heavier hop load. He chose rich caramel malts. "You put crystal malts with those big, fruity hops, that creates a whole new flavor," he said.

Red Ale
Original gravity: 14.5-16.5 °P (1.059-1.067)
Final gravity: 3.0-4.0 °P (1.012-1.016)
IBU: 55-65
ABV: 6.0-6.5%

Grain bill:
75% North American 2-row base malt
10% German Vienna malt
10% 70-80° L Scottish crystal malt
5% 120° L English crystal malt

Mashing:
Add 3.0 oz./bbl $CaSO_4$ and 2.0 oz./bbl CaCl to mash liquor
Infusion mash at 150° F (65.5° C)
Sparge at 164° F (73° C)

Hops:
CTZ, 75 minutes (40 IBU)
Citra, 10 minutes, 0.13 lb./bbl
Simcoe, 10 minutes, 0.13 lb./bbl
Cascade, 10 minutes, 0.25 lb./bbl
Cascade, 0 minutes, 0.5 lb./bbl

Boiling: 90 minutes
Yeast: American Ale, 0.75 million cells per milliliter per Plato
Fermentation: Ferment at 68° F (20° C) and let ride to 74° F (23° C) when two-thirds of fermentation is complete.

Dry hops:
Cascade 1 lb./bbl
Simcoe 0.15 lb./bbl
Citra 0.15 lb./bbl
Dry hop at 65° F (18° C), crash cool on third day for two days, then transfer.
Packaging: 2.5 volumes CO_2 (5 g/L)

Meantime Brewing
London, England

Meantime Brewing founder Alastair Hook grew up in Greenwich, lives in Greenwich, and runs a brewery in Greenwich. "I'm a Londoner. You can't take London out of a Londoner," Hook said. "If you grow up a fan of the Charlton Football Club, your ultimate dream is to go back and play for that club."

Appropriately, 90 percent of the sales of its *London Lager* are in the East End, and all the ingredients, other than yeast, come from within 100 miles of the brewery. The hops are Golding and Fuggle, varieties more often associated with fruitier ales. Before providing this variation on the *London Lager* recipe, head brewer Steve Schmidt, a transplant from America, talked about experimenting with different varieties of Golding that are still grown but not necessarily that easy to find.

"English hops are so, well, English," Hook said. "They are pent up. They are (nicely) subtle, but they aren't for every beer."

Meantime has a ROLEC HOPNIK, using it primarily for late additions and as a hop back. When Schmidt does use it to dose the first addition, wort recirculates through the hops for 10 minutes, then is pumped back into the wort kettle for the rest of the boil. Meantime gets 35 percent utilization on a 75-minute addition, 14 percent on the whirlpool addition, and 12 percent on the cooling addition (essentially a hop back addition done while cooling the wort on the way to the fermenter).

"It's a prettier hoppiness," Hook said, comparing the hop back to dry hopping. "Dry hops are more in your face."

English Lager
Original gravity: 10.7 °P (1.040)
Final gravity: 2.5 °P (1.010)
IBU: 30
ABV: 4.5%

Grain bill:
69% Muntons Pilsner malt
31% Muntons Flagon pale ale malt

Mashing:
66° C (151° F)

Hops:
(All hops whole leaf)
Golding, 6.6 AA, 5 kg in 100 hL, 75 minutes (12 IBU)
Golding, 6.6 AA, 5 kg, whirlpool (5 IBU)
Fuggle, 6.1 AA, 5 kg, whirlpool (5 IBU)
Golding, 6.6 AA, 5 kg, cooling [hop back addition] (4 IBU)
Fuggle, 6.1 AA, 5 kg, cooling [hop back addition] (4 IBU)

Yeast and fermentation: Yeast 34/70, 12° C (54° F) for 10-14 days, and allow yeast to free rise to 15° C (59° F) about 1.5 °P before terminal gravity. Take a sample and force it warm to make certain diaceytl reduced. 5 days at 5° C. Chill to 1° C (34° F). Lager for 2 weeks.
Packaging: 2.3-2.5 volumes CO_2 (4.6-5 g/L)

Pivovar Kout na Šumavě
Czech Republic

Evan Rail, author of *Good Beer Guide Prague and the Czech Republic*, headed into the rural Czech countryside on a February day when the temperature was nearly 40 below (a point where Celsius and Fahrenheit start to meet) to collect this recipe and write a report. Here it is, in his words:

Almost immediately after the brewery in Kout na Šumavě reopened after a 36-year closure in 2006, it quickly earned a name for itself as a producer of excellent 10° and 12° golden lagers, both of which were quickly rated among the best beers in a country that invented, and loves, Pilsner-style half-liters. But beyond the pales, Pivovar Kout na Šumavě also makes two black beauties, both of which exhibit a startling amount of hop character. The 14° is slightly stronger than the average Czech *tmavé pivo*, or "dark beer," which is more often seen at 10° to 13°, while the *Kout 18°* is a monstrous Czech take on the Baltic porter style.

In all of his beers, hlavní sládek—"brewmaster"—Bohuslav Hlavsa uses Saaz hops exclusively, only the true Žatecký poloraný červeňák hops, known locally as ŽPČ, and only those sourced from the original hop region of Žatec.

During a Saaz shortage as the brewery was starting up, Hlavsa initially experimented with other varietals.

"Earlier, we tried to brew with other hops, ones with more hop character and aroma, and we tried newer Czech varieties like Sládek, but it just didn't taste right."

That shortage has since been reversed, with many large breweries using other varietals for bittering, leaving plenty of Saaz available.

"Now there is so much ŽPČ on the market that hop farmers are tearing out the bines," he said. "They're *liquidating* them," he said, using a word with especially stark connotations in Czech.

Kout na Šumavě uses Type 90 pellets, purchasing hops every three or four months and adjusting recipes according to alpha acid content. "Every year it's different, 3 to 3.8," said Hlavsa, "sometimes up to 4 or even 4.2."

Despite the strong hop character of Kout beers—the pale 12°, a true Bohemian Pilsner, clocked 44.2 IBUs in a recent lab analysis—Hlavsa attributes much of the credit to water quality and process.

"The real taste of the beer, of Czech beer, comes from decoction," Hlavsa said. "The hops are just flavorings, like spice in a soup." The brewery's water, sourced from a well in the old-growth Šumava forest, has been tested and rated as *kojenecká voda*, or "nursling water," with very little dissolved minerals.

"Hard water doesn't work well for brewing Czech beers," says Hlavsa.

As for the recipe, Hlavsa has a few comments:

"When beers are stronger, like 14° or 18°, the fermentation is less complete, and they're sweeter, so we can add more hops to compensate. Many brewers like to make a *tmavé pivo* which is just sweet, but we like to make ours bittersweet."

And what is *tmavé pivo*?

"It should have a real darkness, around 80 EBC, but you can still see through it—it should be clear. In terms of aroma, you can really smell the ŽPČ, and it should have a pronounced caramel flavor."

So what's the difference between a Czech *tmavé pivo* and a German *schwarzbier* or *dunkel*?

"I've never tasted a *schwarzbier*," said Hlavsa. "Nor any *dunkels*."

14° Tmavé Speciální Pivo

Original gravity: 14 °P
Final gravity: 5 °P
IBU: 34
ABV: 5.8%

Grain bill:
77% Weyermann Pilsner malt
10% Weyermann Munich II malt
10% Weyermann CaraMunich II malt
3% Weyermann Carafa Type 3 malt

Mashing:
Extremely soft water and a double-decoction mash

Hops:
Saaz (Žatecký polораný červeňák) pellets and a 90-minute boil. "Add one-third of the hops at the start of boil, one-third after 30 minutes, and the final third after another 30 minutes, which is 30 minutes before the end of the boil," said Hlavsa. After boil, Kout uses a whirlpool for 30 minutes before chilling to fermentation temperatures.

Yeast and fermentation: Kout uses the H yeast strain, purchased from Budweiser Budvar. "We pitch the yeast at 8° Celsius (46° F), but fermentation raises the temperature to 10.5 or even 11° Celsius (51-52° F)," said Hlavsa. "When fermentation is 60 percent complete, we transfer it to the lagering tanks, where it ripens for 60 to 90 days, sometimes up to 4 months." Lagering is at 2° C (36° F).

Tip: "It's best to brew dark beer at night," said Hlavsa, "because that way the darkness gets into the beer."

Private Landbrauerei Schönram
Petting/Schönram, Germany

Eric Toft is a member of *Bier-Quer-Denker*, a group of German brewers dedicated to expanding their own beer culture, drawing inspiration from the culture and styles of other countries. Roughly translated, *Bier-Quer-Denker* means "beer lateral thinker." Hans-Peter Drexler at G. Schneider & Sohn is another member.

Toft, a Wyoming native, in fact brews an IPA using American hops (shipping most of it to Italy) but he's also a spokesman for the German Hop Growers' Association. He's vocal about expanding, rather than abandoning, tradition. Since he took over as brewmaster at Private Landbrauerei Schönram in 1998 he has gradually made the recipes his own, increasing hopping rates 10 to 15 percent on average, going against a trend in Germany. Brewery sales have more than doubled, again bucking a national trend.

Schönramer Pils, brewed with lower alpha aroma hops throughout, won medals at the European Beer Star competition in 2009, 2010, and 2011, and the World Beer Cup in 2012. *Schönramer Hell* accounts for more than half of sales and won gold at the European Beer Star in 2012. The recipe here is not quite as assertive as his own, but more so than those brewed closer to Munich.

More than a half-dozen small Bavarian breweries served American-style IPAs at a festival in Munich in the spring of 2012, countering a misconception outside of Germany that its brewing industry is moribund. Pundits put the blame on the *Reinheitsgebot* (the beer purity law that limits ingredients permitted in beer). Toft disagrees and once wrote an essay for a German brewing magazine to make his point.

"There has been a collective, though not all brewers are guilty of this, mass misinterpretation of the *Reinheitsgebot—Reinheitsgebot* as *Einheitsgebot*, meaning all beer must taste the same or all brands are interchangeable. Over the years, processes and technology in the breweries have also become very similar. I see the *Reinheitsgebot* as just the opposite," he said. "Because we are forced to work within these narrow confines, we should see it as motivation for creativity and opportunity to set our brands apart from the others. This begins with the selection of the raw materials and carries through the entire process."

Bavarian Helles
Original gravity: 11.8 °P (1.047)
Final gravity: 2.3 °P (1.009)
IBU: 19
ABV: 5%

Grain bill:
78% Pilsner malt
22% acidulated malt

Mashing:
Decoction mash (see Victory Brewing, pp. 270-271, for details)

Hops:
Hallertau Tradition, 6% AA, 75 minutes (15 IBU)
Hersbrucker, 3% AA, 15 minutes (3 IBU)
Spalter Select, 5% AA, 15 minutes (2 IBU)

Boiling: 90 minutes
Yeast: Weihenstephan 34/70
Fermentation: 9-10° C (48-50° F) until target reached, usually 7 to 8 days. Lager 2 to 3 weeks at 3-4° C (37-39° F), then gradually drop temperature to -1° C (30° F) over 5 days and hold. Total lagering time 5 to 6 weeks.
Packaging: 2.5 volumes CO_2 (5 g/L)

G. Schneider & Sohn
Kelheim, Germany

Brewmaster Hans-Peter Drexler calls *Mein Nelson Sauvin,* which Schneider released in 2011, the culmination of more than 10 years of brewing experiences.

"I had a crucial experience in the year 2000 when I visited the U.S. for the first time. I found pale ales and IPAs with funky and refreshing notes of citrus and grapefruit," he said. American brewers explained Cascade hops contributed those aroma and flavors, and shortly thereafter Drexler began experiments using imported Cascade hops and Schneider's yeast.

He remembered a story brewery owner Georg Schneider VI (who, like his ancestors, is a diploma brewer) told about a special Oktoberfest wheat beer style brewed with a large amount of hops at the Schneider Weisse brewery between 1920 and 1930. The story was, they brewed it at the end of the wheat beer brewing season in April or May. To keep the beer in good condition and safe from infection they used all the hops that remained in their cellars. That beer was called *Wiesen Edel Weisse.*

Drexler's experiments with Cascade culminated in *Georg Schneider's Wiesen Edel Weisse,* a new version of the wheat beer of the 1920s, with 14 °P, 6.2% ABV, and between 25 and 30 IBU. He described it as "a small revolution on the wheat beer market," because it had about twice as many bitterness units as any other wheat beer.

"The second step of inspiration happened few years later in 2007 … (when) Garrett Oliver from the Brooklyn Brewery and I launched *Hopfenweisse,*" he said. "I am sure it was definitely the first dry-hopped wheat beer in Germany, with 40 to 50 IBU." They used Saphir, at the time quite new, for dry hopping.

Drexler next tweaked the recipes for two standards at Schneider. "The idea was to get more freshness and drinkability to these beers. They should taste funky and balanced. It is not only the hops which works in that way. The malty character and the spiciness or fruitiness show the direction to a balanced and funky taste."

He replaced Hallertau Tradition and Magnum in *Blonde Weisse* with 100 percent Saphir and added a bit of Cascade (late hop addition) to *Kristall.* "The results were amazing," he said.

That set the stage for *Mein Nelson Sauvin.* "The idea was to build a bridge from characteristic traditional wheat beer flavors to the wine aroma. (For that) I found Nelson Sauvin hops from New Zealand and yeast from Belgium combined with local wheat and barley malt," he said. It was the first time Schneider used any yeast other than its own.

In deference to the tradition of the Belgian brewery from which it came, Drexler didn't name the source of the yeast. However, other *Bier-Quer-Denker* members have brewed with yeast collected at the Westmalle Trappist brewery.

"In Germany we have a saying: Tradition does not mean keeping the ashes but carrying on the fire," Drexler said. "In that sense hops could help to continue the Bavarian tradition of brewing wheat beer."

Mein Nelson Sauvin

Original gravity: 1.069 (16.8 °P)
Final gravity: 1.013 (3.3 °P)
IBU: 29
ABV: 7.3%

Grain bill:
60% local variety Hermann (6 EBC) wheat malt
20% local variety Marthe (6 EBC) barley malt
20% Urmalz (Munich-style 25 EBC) barley malt

Mashing:
One decoction, targeting high attenuation

Hops:
Hallertau Tradition, 50 minutes (8 IBU)
Nelson Sauvin, 15 minutes (15 IBU)
Nelson Sauvin, 0 minutes (6 IBU)

Boiling: 60 minutes
Yeast: 3 L/hL Schneider yeast from propagation tank. 0.5 L/hL Belgian yeast
Fermentation: 7 days, beginning at 16° C (61° F), allow to rise to 22° C (72° F), reduce to 12° C (54° F)
Bottling: Refermentation in the bottle, using *speise* (unfermented wort). 3.3 volumes CO_2 (6.5 g/L)

Sierra Nevada Brewing
Chico, California

The first time Steve Dresler brewed a wet hop beer at Sierra Nevada (see p. 198) he added five to six times the weight in wet hops that he would have were the hops dry. Soon he increased that to between seven to eight times the dry weight. The simple explanation is that wet hops contain about 80 percent water, while dried cones have about 10 percent.

To brew *HopTime Harvest Ale,* Vinnie Cilurzo at Russian River makes three hop additions, at the beginning of a 90-minute boil, with 30 minutes to go, and at the end. He knows he's on track to hit his target gravity when the reading with 30 minutes matches his final target, because the last two hop additions will contain as much moisture as will evaporate in 30 minutes. He remembers that before he brewed *HopTime* for the first time, Dresler advised him, "Vinnie, don't be intimidated or scared by the amount of hops you are going to add. It seems like a lot, and it is. But in the end, it still probably won't be enough."

Dresler allows it is a fine line. "You want to make sure you aren't overdoing it," he said. "There's a learning curve. You only get one shot a year … it's a long learning curve." He actually gets two shots a year. He brews *Northern Hemisphere Harvest Ale* with wet hops delivered directly from the Yakima Valley, and *Estate HomegrownAle* with organic hops grown on the brewery property.

"One of the challenges is getting them into the liquid," Dresler said. "It's a matter of getting the wort into the oil and vice versa."

That's why Cilurzo warns brewers not to pack wet hops into hop straining bags: It will reduce wort-to-hops contact. He also suggests brewers might be wise to shorten their brew length, because the kettle simply may not hold the additional hops as well as a normal amount of wort.

Northern Hemisphere Harvest Ale

Original gravity: 1.069 (16.6 °P)
Final gravity: 1.016 (4 °P)
IBU: 65 IBU
ABV: 6.7%

Grain bill:
88% 2-row pale malt
12% British caramel malt (65° L)

Mashing:
Calcium 150 ppm, sulfate 300 ppm, chloride 30 ppm
155° F (68° C)

Hops:
(Weights are wet, 8 times dry weight)
Fresh picked wet Cascade, 100 minutes, 2 lbs./bbl
Fresh picked wet Centennial, 100 minutes, 3 lbs./bbl
Fresh picked wet Cascade, 20 minutes, 10 oz./bbl
Fresh picked wet Centennial, 20 minutes, 10 oz./bbl
Fresh picked wet Cascade, 0 minutes, 1 lb./bbl
Fresh picked wet Centennial, 0 minutes, 1 lb./bbl

Boiling: 100 minutes
Yeast: California Ale
Fermentation: Pitch at 62° F (17° C). Allow to rise to 68° F (20° C) and hold 7 days. Rack to secondary, hold at 68° F (20° C) for additional 5 days. Slowly crash chill to 38° F (3° C) on day 14 of fermentation (7 in primary, 7 in secondary).
Packaging: Prime with fresh yeast (California Ale) at bottling, 2.5 volumes CO_2 (5 g/L). Store at 65-68° F (18-20° C) for 10 days for bottle conditioning.

Urban Chestnut Brewing
St. Louis, Missouri

Urban Chestnut Brewing describes its beers as members of a Reverence series or a Revolution series, giving drinkers a good idea of when to expect something traditional and when the beers might break a few rules. Co-founder and brewmaster Florian Kuplent brews *Hopfen* with malts from Germany and Hallertau hops, fermenting it with Weihenstephan 34/70, the workhorse lager yeast commonly used throughout Bavaria, yet it lands firmly on the Revolution side of the menu. He ferments it warmer and dry hops it with German hops and American zeal. Beer rating sites classify it as an "American IPA," but, more accurately, it fits in no style category.

In *The World Atlas of Beer*, Stephen Beamont writes, "IPA-style hopping with German Hallertau hops gives this unique brew a seductively floral aroma and a flavor evocative of a lightly fruity pale ale crossed with aromatic Riesling (p. 192)."

Not long after Urban Chestnut opened in 2011 the brewery hosted its first festival. Hopfenfest featured beers brewed with hops from the Hallertau region. Special guests included representatives of the German hop growers, who were on their way back home after attending a brewing conference, and 2011 Hallertau hop queen Christina Thalmaier. She watched drinkers sampling *Hopfen* spiced with dry hops and without. "I like that the brewers use so much of our hops," she said. She also said she liked *Hopfen* "so much" but allowed such a strong beer was better suited for the evening. "For now (the afternoon) a regular beer is better," she said.

Kuplent also brewed IPAs that were recognizably American in the following months, using new wave varieties such as Calypso and Delta. Most often those sold more briskly than *Hopfen*. "They are so far apart," Kuplent said, talking about hops separated by more than an ocean. "Being able to generate those aromas. You wouldn't be able to get that from a European hop variety."

If Anton Lutz has his way that will change, and chances are Kuplent will use those hops.

Hopfen
Original gravity: 1.058 (14.3 °P)
Final gravity: 1.014 (3.6 °P)
IBU: 55
ABV: 6.2%

Grain bill:
85% imported Pilsner malt
15% imported Munich malt

Mashing:
Mash-in at 131° F (55° C).
Heat to 144° F (62° C).
Rest for 45 minutes.
Heat to 162° F (72° C).
Rest for 45 minutes.
Heat to 172° F (78° C).
Transfer mash
Sparge three times, total sparge water volume to be evenly divided between the sparges.

Hops:
Hallertau high alpha variety (Taurus, Merkur, or Magnum), 90 minutes (25 IBU)
Hallertau Tradition, 60 minutes (15 IBU)
Hallertau Mittelfrüh, 15 minutes (15 IBU)

Boiling: 90 minutes
Yeast: Weihenstephan 34/70 lager. Pitch 15 million cells/mL.
Fermentation: Ferment at 59° F (15° C) until complete, usually 5-6 days. Cool to 30° F (-1° C), and mature for 3 weeks.

Dry hops:
Hallertau Mittelfrüh: 1 lb./bbl
Hallertau Tradition: 1 lb./bbl
Dry hops are added at fermentation temperature, then beer is cooled. Dry hop for 3-5 days.
Packaging: 2.5 volumes CO_2 (5 g/L)

Victory Brewing
Downingtown, Pennsylvania

Early in 2012 Victory Brewing offered further proof of how much difference location may make in the character of what would otherwise seem to be "the same" hop. As Victory has grown, so has its appetite for Tettnanger hops, and brewmaster Ron Barchet must select hops from more farms. After the 2011 harvest he brewed five single-hop Pilsner beers, the same except for the Tettnanger hops used. They came from four different fields. One farmer harvested hops on several different dates, and Barchet took two lots from him.

Victory served the beers, along with a blend of all five, at an event it called *Terroir des Tettnangs* and later at the brewery restaurant, giving drinkers a chance to taste the difference.

The *Kellerpils* recipe Barchet provided also illustrates how a seemingly small change may have a dramatic impact on flavor.

"*Kellerpils* often will hold a tighter grip on the hops than its filtered counterparts. Bitterness substances will attach themselves to the cell walls of the remaining yeast, and, therefore, provide a more bitter beer than its filtered counterpart," Barchet said. "The degree of bitterness can be noticeably higher in beers that are young, with more yeast in suspension. Conversely, longer aged *Kellerpils* will have very few cells in suspension, adding scant more bitterness than a filtered *Pils*. Flavor and aroma are also enhanced to a lesser extent in a *Kellerpils*."

Kellerpils

Original gravity: 11-12 °P (1.044-1.048)
Final gravity: 1.6-2.2 °P (1.006-1.009)
IBU: 35-45
ABV: 4.8-5.2%

Grain bill:
97-100% German 2-row Pilsner malt, medium modification
0-3% acid malt, if needed, to hit mash pH of 5.2-5.3

Mashing:
Single decoction:
Mash-in entire grist at 122° F (50° C), using a thin mash.

Immediately raise to 144° F (62° C).

If you have a separate mash kettle:

Move 25% of the mash to the mash kettle.

Hold for 10 minutes, then raise 2° C per minute to boil.

Boil 1-3 minutes, then move back into the mash tun.

This should bring the main entire mash up to 70-72° C (158-162° F).

Hold at 70-72° C (158-162° F) for 15 minutes, then raise to 77° C (171° F).

If you do not have a separate mash kettle, but can move mash to and from lauter tun:

Move 75% of the mash to the lauter tun, which should be preheated by foundation water.

In the mash tun, hold the remaining 25% for 10 minutes; then raise 2° C per minute to boil.

Boil 1-3 minutes, then pump the 75% remaining in lauter tun back into the mash tun.

This should bring the main entire mash up to 70-72° C (158-162° F).

Hold at 70-72° C (158-162° F) until iodine starch check passes, but for a minimum of 15 minutes.

Raise to 77° C (171° F) for mash off, and send entire mash to lauter tun for lautering.

Hops:

Saaz, 60 minutes (5 IBU)

Hallertau Mittelfrüh, 60 minutes (5 IBU)

Tettnang Tettnanger, 30 minutes (10 IBU)

Spalt Spalter, 10 minutes (10 IBU)

Saaz, 5 minutes (5 IBU)

Sládek, 5 minutes (5 IBU)

The hop timings are based on whole flower hops. If using pellets, shifting the hop additions later in the boil would yield similar results.

Boiling: Boil as long as necessary to achieve proper reduction in DMS precursors, but not longer. Depending on the veracity of the boil and its capability to reduce DMS precursor, the length of boil usually is 65-90 minutes. Boiling longer than necessary will increase color and decrease flavor stability.

Yeast: Weihenstephan 34/70, pitched at 15 million cells per milliliter at 10° C (50° F); aerate to 8 ppm O_2.

Fermentation: Ferment at 10° C (50° F) until 5 °P (1.010), then cool slowly (1° C/day) to 4° C (39° F). Hold at 4° C (39° F) until fermented and when diacetyl is no longer noticeable. Cool to 0° C (32° F) and hold for 2-3 weeks. Racking the beer into a lager tank after fermentation is complete improves final beer taste and clarity.

Packaging: 2-2.4 volumes CO_2 (4-4.8 g/L). The wide range covers the variant of a *Kellerpils* called *ungespundet*, which means "not bunged" in German. The resulting low dissolved CO_2 content makes for a creamier, softer take on the *Kellerpils* style.

Weyerbacher Brewing
Easton, Pennsylvania

In 1998 Weyerbacher Brewing in Pennsylvania introduced a beer made with a blend of seven hops that was, for the time, shockingly aromatic and bitter. A dozen years later founder Dan Weirback spoke candidly about the recipe for *Hops Infusion India Pale Ale*. "I have a problem with smell. I have a problem getting hops in some beers I get on draft," he said. "I can't really pick out most of the different hops in our beers. That's part of the reason there are so many hops in *Infusion*."

He and head brewer Chris Wilson reformulated the recipe in 2006. "It needed to come up to speed with the rest of the industry," Weirback said. "It had become way underhoppy, although it was the same beer." In the process, when they started looking for new hop varieties, Simcoe caught their attention because of its low amount of cohumulone and high amount of oil. It ended up being the centerpiece of *Double Simcoe IPA*, a 9 percent ABV beer that directed attention to the hop at a time when it was not nearly as popular as it is today. It is Weyerbacher's third-best seller, beyond a Belgian-inspired *tripel* and a barley wine.

Verboten (German for "forbidden") is something altogether different. It was the first beer Wilson brewed as part of a series of one-offs, called *Alpha* because the plan was to work through the NATO alphabet. "We were surprised. We were going bigger and bigger, and this beer wasn't that strong," Wilson said. "But it sure had a lot of flavor, something our customers were looking for."

Weyerbacher first called the beer *Zotten* (Dutch for "fools") when it released it as a seasonal, but after a Belgian brewery sued because it made a beer with a similar name, Weirback renamed it *Verboten*.

The pairing of Belgian yeast strains and American hop varieties, sometimes in beers that include large quantities of American hops, has resulted in numerous very popular beers since Brasserie d'Achouffe released *Houblon Dobbelen IPA Tripel* in 2006. However, not all brewers or drinkers agree that it's a good combination. *Verboten*, which balances fruity esters and a bit of spicy phenolic yeast character with juicy, citrusy aromas and flavor with balanced restraint, makes it hard to argue the combination should be forbidden. .

Verboten

Original gravity: 15 °P (1.061)
Final gravity: 4.0 °P (1.016)
IBU: 45
ABV: 5.9%

Grain bill:
75% Muntons pale malt
15% wheat malt
10% CaraHell malt (or other light-colored crystal malt)

Mashing:
149° F (65° C) 60 minutes

Hops:
Centennial, 60 minutes, (45 IBU)
Cascade pellets, 0 min (added before whirlpool), 0.5 lb./bbl
Cascade whole hops, 0 min (added to hop back), 0.5 lb./bbl

Boiling: 60 minutes
Yeast: Wyeast 1214
Fermentation: Six days at 72° F (22° C). Let stand for 48 hours. Lower temperature to 58° F (14.5° C) to 48° F (9° C) to 32° F (0° C) next 48 hours. Hold for 6 days.
Packaging: Bottle condition with primary yeast. 2.9 volumes CO_2 (5.8 g/L)

11

Epilogue

The future has already arrived, so what about the future?

Ralph Olson has a sweeping view of a good portion of the Yakima Valley from his back patio on a rise at the outer reaches of Yakima itself, Moxee distant to the left, the Cascade Mountains to the right. He sees the rest in his mind's eye, talking about the areas with sandier soil, the regions where hops yield greater amounts of alpha acids, or commenting on the harvest-related skills of individual farmers.

He retired as CEO of Hopunion in 2011 after 35 years of selling hops, easing his way out over a few harvests (he and his wife continue to ship rhizomes through the mail each spring). In 2000 the Brewers Association gave Olson and Ralph Woodall—not surprisingly, many in the industry simply call them "the Ralphs"—its annual Recognition Award for their work at Hopunion. Today all of Yakima's hop merchants have established a strong connection with smaller breweries, as well as homebrewers. In the early 1980s that was different.

"We were mainly dealing with Anheuser-Busch, Coors, Stroh's, Miller, Pabst, Olympia, and Rainier," Woodall said. He was standing out of the way as brewers, mostly from rather small companies, crowded around Hopunion's booth at the 2010 Craft Brewers Conference, some asking the most basic of questions, others very sophisticated ones about new varieties.

With the harvest nearing conclusion in 2011, Olson sat on the patio with a hop-forward beer in hand. One story he told—he has at least one for about almost every variety—illustrates how recently "special" aromas became special.

A cross made in 1974 produced a seedling called W415-90 that advanced to Hop Research Council field trials by the mid-1980s. "We're at a meeting with hop growers, talking about if we should increase acreage of W415-90," Olson said. No major breweries had shown interest in the variety. "I stood up and said, 'I hate to see us get rid of it. Some craft brewers were interested in it.' Another dealer asked, 'Are they paying for the program?' "

The growers voted to discontinue trials for W415-90. "I went to the farmer who had it and told him we'd change the name and keep growing them," Olson said. He gave the variety a new name, CFJ-90, and kept selling it. "I started to get nervous; it was getting more popular. We got to about 10 acres, and I had to go to Prosser (the USDA hops research station in Washington) and confess. I told them (specifically, Steve Kenny) I didn't know what to do. He laughed and said, 'Let's give the bastard a name.' "

Because the state of Washington celebrated 100 years of statehood in 1989 they chose the name Centennial. As recently as 2005, growers still planted little more than 100 acres of Centennial. The demand since reflects the growth in blossoming sales of IPA (most drinkers simply use the initials rather than asking for "India pale ale"). Its aroma is unique but not universally appealing. "The flavor of flavor hops, these up to now were regarded in our program as not suitable for brewing," Zdeněk Rosa, manager of Hop Growers Union of the Czech Republic, said, candidly. "These are off-flavors for us."

Today Centennial is a key component in such fast growing brands as *Lagunitas IPA, Bell's Two-Hearted Ale,* and Bear Republic's *Racer 5.* Between 2010 and 2011 farmers in Washington, where most of it is grown, boosted Centennial acreage from 357 to 641 and production from 639,400 pounds in 2010 to 899,400, an increase of 40 percent that almost perfectly mirrors 41 percent more IPA sales in 2011 than in 2010.

Nobody forecast such growth, and nobody was talking with confidence about how long it might continue. As recently as the beginning of 2008 pale ale was the best selling craft beer style in supermarkets, followed by amber ale, amber lager, wheat beers, and then IPA. In the four years between the end of 2007 and end of 2011 sales of IPA increased 260 percent as it became the No. 1 craft style. A bit of unattributed commentary, obviously from an Englishman, that appeared in *Hop Culture in the United States* in 1883 provides a reminder that extrapolating from that data could be deceiving.

"The brewing industry is not exempt from the influence of fashion. A careful survey of the types and descriptions of beers in vogue at different times will show that fashion has had something to do with our trade," the author wrote. He described changes in beer dating to back before hops became an essential ingredient, and considered what might be next in England. "We will not further refer to the threatened introduction of lager beer into this country, than to say fashion takes strange freaks, and it will be well for brewers to be prepared for all eventualities."[1]

These new fangled hops remain fashionable within what is in fact a niche but perhaps a large enough one to have changed the conversation. Consider this bit of math. In 2011 American beer production shrank by about 4.6 million barrels overall, although craft beer production grew by 1.3 million barrels. A survey of Brewers Association members indicates they use about one pound of hops per barrel, so they would have needed at least an additional 1.3 million pounds of hops in 2011, probably more taking dry-hopped IPAs into account. Production of all beers not classified as craft actually fell 5.9 million barrels. If brewers used two ounces of hops in each of those barrels—an educated guess—that would not have amounted to 750,000 pounds of hops. Despite the hefty drop in beer production overall, American hop usage increased.

Just five years ago the hop supply chain looked much different. In 2007 a worldwide shortage of hops sent prices skyrocketing, the first dramatic increase in more than 30 years, when disease in European fields was a factor in the release of Cascade. The following year brewing giants Anheuser-Busch and InBev merged, and A-B began eliminating or reducing many of its forward contracts.

A-B InBev offered generous terms, basically paying farmers not to grow hops or reducing amounts under contract on a gradual basis. German farmers cut Hallertau Mittelfrüh production in half within a year. In Oregon and Washington Willamette acreage shrunk from 7,257 in 2008 to only 1,256 in 2012.

A Single Hop or 2,012?

Single-hopped beers aren't hard to find any more.

In England, Marston's brewed a series of single-hop beers, offering one a month throughout 2012. The brewery used hops from nine different countries, including both well established landrace varieties such as Saaz and the newly minted Wai-iti from New Zealand.

In 2011 Mikkeller, the Danish "gypsy" brewery, released 19 single-hop beers. Each was made with the same weight of hops, so that those brewed with low alpha varieties such as Hallertau Mittelfrüh were relatively delicate, while those made with high alpha hops provided a lesson in bitterness as well as aroma.

The list goes on, for many drinkers including one from their local brewery or brewpub.

At the opposite end of the spectrum, the Great Yorkshire Brewery in England brewed a beer called *Top of the Hops 2012* with 2,012 varieties, using those that failed in trials at Wye Hops Limited. Brewery director Joanne Taylor said the brewery wanted to support Peter Darby's research. "We couldn't donate money, but we thought a great way we could work together would be to buy those hops from him and brew a fascinating beer," she said. The mix included dwarf varieties, aphid-resistant types, plants with Russian and South African pedigrees, and hops derived from Fuggle and other British varieties.

"We went cold turkey," said John Segal at the Segal Ranch on the eastern edge of Yakima County. His late father, also named John, had begun growing hops for A-B 30 years before and also kept an experimental plot for them. Segal chose to take a buyout and let his Willamette acres lay dormant. When he attended the 2009 Craft Brewers Conference he didn't know anybody at a craft brewery other than Anchor Brewing—his father first sold Cascade to Anchor in 1974—but committed himself to finding craft clients.

He talks about pretty much the same things as craft brewers do. The back of Segal Ranch hats read "All hops are good. Some are better." Signs that say, "Remember this is a food product" hung around the property long before the Hop Quality Group began giving similar ones to other farmers. He emphasizes the care with which his staff dries hops and the amount of oil in them. "It all comes down to the rub," he said. Segal planted 83 acres of hops in 2009, 190 in 2010, 290 in 2011, and 390 in 2012, putting his property at capacity.

The shift from alpha-rich hops to aroma-focused hops continued at a faster pace in 2012 than even Alex Barth predicted at the beginning of the year. Acreage dedicated to hops that American brewers use later in the boil or post-fermentation (those such as Simcoe and Chinook) grew from 32 percent of acres strung in 2011 to 40 percent in 2012. This did not only include high demand, proprietary hops with *aromas du jour:* Centennial acreage was up 93 percent and Chinook 73 percent, for instance.

Growers understand what hop varieties brewers want better than they did just a few years ago—in part, because more brewing companies learned the importance of forward contracts, and in part, because of improved communications (p. 228). Hop shortages in 2007 and 2008 resulted from a combination of circumstances, including natural disasters and poor worldwide inventory control caused by rapid industry consolidation. They also arose because many breweries did not contract for the hops they knew they would need, choosing to buy them on the "spot" market. Farmers planted fewer acres because prices were lower. Surpluses shrank. It didn't take much to trigger a shortage. Prices skyrocketed for those without contracts. Not surprisingly, breweries reacted by contracting for future deliveries. Those new to the process learned that contracting is not all that simple, but there are some key elements: a) it is easier to quickly add acreage of established varieties, b) contracts don't guarantee the lowest prices, but c) they do ensure delivery, with rare exceptions.

New Zealand Hops representative Doug Donelan pointed to the variety Nelson Sauvin as an example. "We grew 90 metric tons (less than 200,000 pounds and not half of Citra production in 2011), all of which was forward sold," he wrote in an email. "Brewers attempting to buy high demand hops on spot markets need to rethink their purchasing strategy. Nelson Sauvin has been in volume growth for the past few years and will continue for the foreseeable future. Hops aren't something you just turn on with a tap." For farmers to plan ahead, brewers must plan ahead as well.

"If you stick with the old way of doing things, the world will go past you. Drinkers in 1947 were different," said John Keeling, brewing director at Fuller's in London. Fuller's uses more than its share of old-line English varieties, particularly Golding, but in 2011 also brewed a beer with Citra and Nelson Sauvin hops, a small run sold only in its own pubs that sold out almost immediately. The brewery used American Liberty, Willamette, Cascade, and Chinook in the more widely distributed and dry-hopped *Wild River* pale ale in 2012.

"The hops fit the market, for different beers for different occasions," Keeling said. "Beer has become for more than going down to the pub and having three pints. You can't make this variety of beers simply with Golding."

The choices don't necessarily need to be the "flavor of the month." Fuller's *Discovery*, brewed with malted wheat as well as barley, is hopped with Hallertau Mittelfrüh offspring Liberty and Saaz. "Hops reflect what's going on in craft beer," Keeling said. "The difference that malt makes is negligible. Is there a difference? Can anybody tell the difference? It is very easy for (drinkers) to see the difference between Golding and Cascade."

The difference between Saaz grown in the Czech Republic and any other hop, even those closely linked genetically, is equally apparent. "We like to say there is no replacement for the Saaz hop," said Rosa, who is also chairman of the Bohemia Hop Company. "It is all influenced by the climate, the local soil, the length of day ... "

Dan Carey at New Glarus Brewing describes the landrace varieties, specifically Saaz and Mittelfrüh, rather succinctly. "They taste like beer. When you use them in beer they taste like beer," he said.

The Saaz hop itself accounts for about 83 percent of Czech production, and in 2011 the country grew more hops classified as "aroma" than in all the United States. "Saaz has been around for 1,000 years, so it still works," Rosa said. He lifted his left hand, palm up, to nearly shoulder height, as if he were holding a single cone. "We have something that is perfect," he said, without a touch of arrogance.

Nonetheless, in a changing world the Czechs are changing, releasing new varieties, working with Peter Darby from the U.K. to develop varieties suitable for growing on low trellises, mapping molecular markers, and finding wild hops useful as genetic resources. As with hop scientists everywhere they are also focused on discovering nonbrewing uses for hops.

Twenty years ago 70 percent of Czech hops were sold whole, and now only 4 percent are. Half the pellets are Type 90 and half Type 45, with a very small amount of Agnus sent to Germany for extraction. About 135 farmers remain, 101 of them members of the growers union, half of them small, and half with farms ranging from 150 acres to 750 acres.

Rosa's father grew hops, and his grandfather both farmed and was involved in research. As a youth he was in charge of the hop kiln during harvest. "I know all sides of the family business," he said. He remembers when hundreds of outsiders would appear in his village. "Hop picking turned it upside down," he said. "Today you notice the tractor, but you don't notice new people."

The summer of 2011 was wet throughout Žatec, delaying the harvest of other crops, leaving farmers who grow a range of crops behind when it came time to pick hops. Rosa had been talking to farmers on this particular morning, telling them they needed to get into whatever hop fields weren't too muddy. Josef Vostřel, head of plant protection, explained alphas levels were good and downy mildew would soon be a threat.

No matter how many ways scientists find to analyze the hop, it's still an agricultural product. Only a tiny portion of beer drinkers will ever see a hop plant, and then it may be among only a few decorative bines at a brewery. But those who work with hops daily, including brewers as well as farmers, are serious when they talk about being "scratched by the hops." Three more examples:

- It is not necessarily clear what the hops in a small yard at the Monastery of Christ in the Desert north of Abiquiu, New Mexico, may come to. They include several varieties grown from rhizomes acquired from Todd Bates (p. 74) near Embudo, presumably American wild hops. The monks contract to have *Monks' Ale*, *Monks' Wit*, and *Monks' Tripel* made at a nearby brewery and sell it in several states. Only the tripel has been brewed with monastery hops, but the monks built a small, solar-powered brewhouse on their property, and the hops may end up in beers made there.

Both Abbot Philip, the abbot for the monastery, and Brother Christian, who represents the brewery on public occasions, began their monastic lives at Mount Angel in Oregon. They talk fondly of baling hops in the fields that monastery overlooks. Abbot Philip shared his thoughts in an email:

"I entered the monastery in 1964, in August. My first exposure with hops was in the summer of 1965. I was one of the monks working in the post-harvest processing of the hops. All of us younger monks were expected to work there. I had no idea what hops were at that time, but I knew that they produced a good income for the monastery that was used to help educate many of us for the priesthood.

"The next year, in 1966, I was in charge of the hop crews, because I was able to get the younger monks to cooperate and actually to have a really good time baling hops. It was hot work and needed a lot of energy. The faster we worked, the sooner we got done.

"We had what was then a state-of-the-art baling machine. Two monks would work upstairs, pushing hops into the baler with a large piece of plywood. Every day we would have to estimate again how many hops would be needed to make a 200-pound bale. The difference each day was because of the different degrees of humidity in the hops.

"I was in charge of the hop baling crews until 1973. I loved that kind of work, because it required huge amounts of energy.

"By the end of the hop baling season those of us who had worked there were really tired but generally happy that we had done a good job. Our clothes were totally destroyed and useless, and so we simply put them in the garbage at the end of the season.

"This was a wonderful work for me as a young monk and is one of the great memories of my early monastic life."

- As the spring of 2012 arrived outside of Moxee, Kevin and Meghann Quinn and Meghann's younger brother, Kevin Smith, began taking down trellises in Field 41 of Loftus Ranches to make room for their Bale Breaker Brewing Company. "They've already told us we won't get any special deals on hops," Meghann Quinn said, grinning and nudging the shoulder of another brother, Patrick Smith, now vice president of Loftus Ranches.

Their father, Mike Smith, is a third-generation hop farmer. "It was about eight years ago, my husband and I were just dating, and I pitched the idea to my dad," Meghann Quinn said. "He said, 'That's the worst idea I ever heard.'" Times and opportunities change. The Yakima Valley is thick with vineyards and wine tasting rooms. A brewery surrounded by hop fields fits right in.

As its contracts with Anheuser-Busch InBev and for high alpha varieties expired, Loftus planted more hops of interest to craft brewers. "We're headed toward 80 percent for craft," Patrick Smith said. "Is it a risky strategy? We're already tied to this industry." Even before the

Hop Quality Group made plans to study the effects of reducing kilning temperatures, Smith conducted his own experiments and invited brewers to assess the results.

- On the Bentele farm outside of Tettnang, Georg Bentele and his family grow about 55 acres of hops, 70 percent of them Tettnanger. Bentele's son brews the beer in the 12-hectoliter (about 10-U.S.-barrel) brewhouse on the grounds. Bentele's daughter, who was once a Tettnang hop queen, was the first brewer.

The beef and pork the Benteles serve in their restaurant and beer garden, *Brauereigasthof Schöre*, comes from animals they raise. They distill eight flavors of schnapps, some with apples they also grow. The three lagers (there is also a *weissbier*) are not filtered but age a minimum of two months, most for three, and pour sparkling bright. The hops, of course, are from the surrounding fields.

Bentele, whose grandfather first planted a single hectare of hops in 1906, sells his hops directly to brewers, most of them to Victory Brewing in Pennsylvania. He planted three acres of Perle in 2008 and said that was a mistake. He explained he thought farmers in Tettnang should grow only hops with the finest aroma—at the Bentele farm this also includes a little Mittelfrüh—because that is what makes the region special.

"We like to say we live with the hops," said Jürgen Weishaupt, managing director of *Tettnanger Hopfen*. Like Fritz Tauscher of the *Kronen-Brauerei* he attended the 2011 Craft Brewers Conference and was astonished by the interest in all hop varieties. "U.S. brewers are hop freaks."

Christina Thalmaier, the 2010 Hallertau hop queen, was equally impressed. "The brewers taste and smell and go 'hmmm, there is this variety of hops in the beer.' Then they talk about it some more," she said. "And then I see people (drinkers), they say, 'It is good' or 'It is not good.' "

David Grinnell of Boston Beer likes eavesdropping on those conversations. "It is positive people are talking about hops," he said. "There are people who have given their life to this plant. I've been known to miss flights. People ask, 'What's with hops?' I don't have a short version of this story."

Nobody does. One word that keeps coming up is synergy. It dashes any attempt to simplify what hops contribute to beer, the how and the why.

"You can't just look at single fractions," said James Ottolini of the Saint Louis Brewery. "That's like listening to one note by itself. I want to know what the orchestra will sound like."

It turns out John Levesque may not have had it completely figured out in 1836 when he summarized the qualities of the hop. He wrote, "Hops are hot, and in the third degree, inciting, abstersive, subastringent, digestive, discussive, diuretic, stomachic, and sudorific—indeed, the spirit of the hop is truly cordial."[2]

Agreed, the hop may be cordial. But when it comes to understanding her, well … we sure ain't there yet.

Notes

1. Ezra Meeker, *Hop Culture in the United States* (Puyallup, Washington Territory: E. Meeker & Co., 1883), 145.

2. John Levesque, *The Art of Brewing and Fermenting* (London: Thomas Hurst. 1836), 145-146.

Bibliography

Arnold, John P. *Origin and History of Beer and Brewing From Prehistoric Times to the Beginning of Brewing Science and Technology.* Chicago: Alumni Association of the Wahl-Henius Institute of Fermentology, 1911. Reprint, *BeerBooks.com,* 2005.

Bailey, B., C. Schönberger, G. Drexler, A. Gahr, R. Newman, M. Pöschl, and E. Geiger. "The Influence of Hop Harvest Date on Hop Aroma in Dry-hopped Beers," *Master Brewers Association of the Americas Technical Quarterly* 46, no. 2 (2009), doi:10.1094 /TQ-46-2-0409-01.

Bamforth, Charles, ed. *Brewing: New Technologies.* Cambridge, England: Woodhead Publishing Limited, 2006.

Barth, H.J., C. Klinke, and C. Schmidt. *The Hop Atlas: The History and Geography of the Cultivated Plant.* Nuremberg, Germany: Joh. Barth & Sohn, 1994.

Beatson, R., and T. Inglis. "Development of Aroma Hop Cultivars in New Zealand," *Journal of the Institute of Brewing* 105, no. 5 (1999), 382-385.

Bennett, Judith M. *Ale, Beer, and Brewsters in England*. New York: Oxford University Press, 1996.

Bickerdyke, John. *The Curiosities of Ale & Beer*. London: Swan Sonnenschein & Co., 1889. Reprint, *BeerBooks.com*, 2005.

Borremans, Y., F. Van Opstaele, A. Van Holle, J. Van Nieuwenhove, B. Jaskula-Goiris, J. De Clippeleer, D. Naudts, D. De Keukeleire, L. De Cooman, and G. Aerts. "Analytical and Sensory Assessment of the Flavour Stability Impact of Dry-hopping in Single-hop Beers." Poster presented at Tenth Trends in Brewing, Ghent, Belgium, 2012.

Buck, Linda. "Unraveling the Sense of Smell (Nobel lecture)." *Angewandte Chemie* (international edition) 44 (2005), 6128-6140.

Buhner, Stephen Harrod. *Sacred and Herbal Healing Beers*. Boulder, Colo.: Brewers Publications, 1998.

Burr, Chandler. *The Emperor of Scent*. New York: Random House, 2002.

Chadwick, William. *A Practical Treatise on Brewing*. London: Whitaker & Co., 1835.

Chapman, Alfred. *The Hop and Its Constituents*. London: The Brewing Trade Review, 1905.

Clinch, George. *English Hops*. London: McCorquodale & Co., 1919.

Cook, Kim. "Who Produced Fuggle's Hops?" *Brewery History* 130 (2009).

Coppinger, Joseph. *The American Practical Brewer and Tanner*. New York: Van Winkle and Wiley, 1815. Reprint, *BeerBooks.com*, 2007.

Cornell, Martyn. *Amber, Gold, and Black: The History of Britain's Great Beers*. London: The History Press, 2010.

_____. *Beer: The Story of the Pint*. London: Headline Book Publishing, 2003.

Corran, H.S. *A History of Brewing*. London: David & Charles, 1975.

Culpeper, Nicholas. *The English Physician*. Cornil, England: Peter Cole, 1652.

Darby, Peter. "The History of Hop Breeding and Development." *Brewery History* 121 (2005), 94-112.

_____. "Hop Growing in England in the Twenty-First Century," *Journal of the Royal Agricultural Society of England* 165 (2004). Available at *www.rase.org.uk/what-we-do/publications/journal/2004/08-67228849.pdf*.

_____. "The UK Hop Breeding Programme: A New Site and New Objectives." Proceedings of the Scientific Commission, International Hop Growers' Convention, Tettnang, Germany, 2008, 10-14.

Darwin, Charles. *The Movements and Habits of Climbing Plants*. London: John Murray, 1906.

Dick, Ross. "Blatz to Offer a New Beer," *Milwaukee Journal,* Aug. 15, 1955, 10.

De Keukeleire, Denis. "Fundamentals of Beer and Hop Chemistry." *Quimica Nova* 23, vol. 1 (2000), 108-112.

Deneire, Bertin. "The Hoppiest Days of My Life." Oral history, 2011, kept at the HopMuseum in Poperinge, Belgium. Available at *www.hopmuseum.be/images/filelib/hopstory.pdf*.

Drexler, G., B. Bailey, C. Schönberger, A. Gahr, R. Newman, M. Pöschl, and E. Geiger. "The Influence of Harvest Date on Dry-hopped Beers," *Master Brewers Association of the Americas Technical Quarterly* 47, no. 1 (2010), doi:10.1094 /TQ-47-1-0219-01.

Ellison, Sarah. "After Making Beer Ever Lighter, Anheuser Faces a New Palate," *Wall Street Journal*, April 26, 2006.

Fink, Henry. "The Gastronomic Value of Odours," *The Contemporary Review* 50, November 1886.

Fisher, Joe and Dennis. *The Homebrewer's Garden*. Pownal, Vt.: Storey Communications, 1998.

Fleischer, R., C. Horemann, A. Schwekendiek, C. Kling, and G. Weber. "AFLP Fingerprint in Hop: Analysis of the Genetic Variability of the Tettnang Variety," *Genetic Resources and Crop Evolution* 51 (2004), 211-220.

Flint, Daniel. *Hop Culture in California*. Farmers' Bulletin No. 115. Washington, D.C: U.S. Department of Agriculture, 1900.

Forster, A., M. Bedl, B. Engelhard, A. Gahr, A. Lutz, W. Mitter, R. Schmidt, C. Schönberger. *Hopfen—vom Anbau bis zum Bier*. Nuremberg, Germany: Hans Carl, 2012.

Forster, Adrian. "What Happens to Hop Pellets During Unexpected Warm Phases?" *Brauwelt International* 2003/1, 43-46.

_____. "The Quality Chain From Hops to Hop Products." Presentation at the 48th International Hop Growers Convention Congress, Canterbury, England, 2001.

Fritsch, Annette, "Hop Bittering Compounds and Their Impact on Peak Bitterness on Lager Beer." Master's thesis. Oregon State University, 2007.

Gent, D., J. Barbour, A. Dreves, D. James, R. Parker, and D. Walsh, eds. *Field Guide for Integrated Pest Management in Hops*. 2nd edition, 2010. Available at *hops.wsu.edu*.

Gilbert, Avery. *What the Nose Knows: The Science of Scent in Everyday Life*. New York: Crown Publishing, 2008.

Gimble, L., R. Romanko, B. Schwartz, and H. Eisman. *Steiner's Guide to American Hops*. Printed in United States: S.S. Steiner, 1973.

Glaser, Gregg. "The Late, Great Ballantine," *Modern Brewery Age*, March 2000.

Grant, Bert, with Robert Spector. *The Ale Master*. Seattle: Sasquatch Books, 1998.

Green, Colin. "Comparison of Tettnanger, Saaz, Hallertau, and Fuggle Hops Grown in the USA, Australia, and Europe," *Journal of the Institute of Brewing* 103, no. 4 (1997), 239-243.

Gros, J., S. Nizet, and S. Collin. "Occurrence of Odorant Polyfunctional Thiols in the Super Alpha Tomahawk Hop Cultivar. Comparison With the Thiol-rich Nelson Sauvin Bitter Variety," *Journal of Agricultural and Food Chemistry* 59, issue 16 (2011), 8853-8865.

Hall, Michael. "What's Your IBU?" *Zymurgy*, Special 1997, 54-67.

Hanke, Stefan. "Linalool—A Key Contributor to Hop Aroma," *Master Brewers Association of the Americas—Global Emerging Issues*, November 2009.

Haseleu, G., A. Lagermann, A. Stephan, D. Intelmann, A. Dunkel, and T. Hofmann. "Quantitative Sensomics Profiling of Hop-Derived Bitter Compounds Throughout a Full-Scale Beer Manufacturing Process," *Journal of Agricultural and Food Chemistry* 58, issue 13 (2010), 7930-7939.

Haunold, Al, and G.B. Nickerson, "Development of a Hop With European Aroma Characteristics," *Journal of the American Society of Brewing Chemists* 45 (1987) 146-151.

Havig, Van. "Maximizing Hop Aroma and Flavor Through Process Variables," *Master Brewers Association of the Americas Technical Quarterly* 47, no. 2 (2009), doi:10.1094/TQ-47-2-0623-01.

Hayes, J., M. Wallace, V. Knopik, D. Herbstman, L. Bartoshuk, and V. Duffy. "Allelic Variation in TAS2R Bitter Receptor Genes Associates with Variation in Sensations From and Ingestive Behaviors Toward Common Bitter Beverages in Adults," *Chemical Senses* 36, vol. 3 (2011), 311-319.

Henning, John. "USDA-ARS Hop Breeding and Genetics Program." Presentation at Winter Hops Conference, Stowe, Vt., 2011. Hertfordshire Federation of Women's Institutes. *Hertfordshire Within Living Memory.* Newbury, Berkshire, England: Countryside Books, 1993.

Herz, Rachel. *The Scent of Desire: Discovering Our Enigmatic Sense of Smell.* New York: Harper Perennial, 2008.

Hofmann, T. "The (In)stability of the Beer's Bitter Taste." Presentation at the 32nd EBC Congress, Hamburg, Germany, 2009.

Hornsey, Ian. *A History of Beer and Brewing.* Cambridge, England: Royal Society of Chemistry, 2003.

Hughes, E. *A Treatise on the Brewing of Beer.* Uxbridge, England: E. Hughes, 1796.

Intelmann, D., C. Batram, C. Kuhn, G. Haseleu, W. Meyerhof, and T. Hofmann. "Three TAS2R Bitter Taste Receptors Mediate the Psychophysical Responses to Bitter Compounds of Hops (*Humulus lupulus* L.) and Beer," *Chemical Perception* 2 (2009), 118-132.

Jackson, Michael. "The Glass of '93 Blossoms Early," *The Beer Hunter,* Oct. 1, 1997. Available at *www.beerhunter.com/documents/19133-000114.html.*

Joh. Barth & Sohn. *The Barth Report.* Nuremberg, Germany:

Joh. Barth & Sohn. Issues accessed from 1911 to current. *www.barthhaasgroup.com/index.php?option=com_content&task=view&id =28&Itemid=30*

_____. *The Hop Aroma Compendium*, vol. 1. Nuremberg, Germany: Joh. Barth & Sohn. 2012.

Kajiura, H., B.J. Cowart, and G.K. Beauchamp. "Early Developmental Change in Bitter Taste Responses in Human Infants," *Developmental Psychobiology* 25, issue 5 (1992), 375-386.

Kaneda, H., H. Kojima, and J. Watari. "Novel Psychological and Neuro-physical Significance of Beer Aroma, Part I: Measurement of Changes in Human Emotions During the Smelling of Hop and Ester Aromas Using a Measurement System for Brainwaves," *Journal of the American Society of Brewing Chemists* 69, no. 2 (2011), 67-74.

Kaneda, H., H. Kojima, and J. Watari. "Novel Psychological and Neu-rophysical Significance of Beer Aroma, Part II: Effects of Beer Aromas on Brainwaves Related to Changes in Human Emotions," *Journal of the American Society of Brewing Chemists* 69 no. 2 (2011), 75 80.

Keller, A., and L.B. Vosshall. "Human Olfactory Psychophysics," *Current Biology* 14, no. 20 (2004), 875-878.

Keese, G. Pomeroy. "A Glass of Beer," *Harper's New Monthly Magazine* 425 (October 1885), 666-683.

Kishimoto, Toru. "Hop-Derived Odorants Contributing to the Aroma Characteristics of Beer." Doctoral dissertation. Kyoto University, 2008.

Kiyoshi, T., Y. Itoga, K. Koie, T. Kosugi, M. Shimase, Y. Katayama, Y. Na-kayama, and J. Watari. "The Contribution of Geraniol to the Citrus Flavor of Beer: Synergy of Geraniol and ß-Citronellol Under Coexistence with Excess Linalool," *Journal of the Institute of Brewing* 116, no. 3 (2010), 251-260.

Kiyoshi, T., M. Degueil, S. Shinkaruk, C. Thibon, K. Maeda, K. Ito, B. Bennetau, D. Dubourdieu, and T. Tominaga. "Identification and Characteristics of New Volatile Thiols Derived From the Hop (*Humulus lupulus* L.) Cultivar Nelson Sauvin," *Journal of Agricultural and Food Chemistry* 57, issue 6 (2009), 2493-2502.

Kollmannsberger, H., M. Biendl, and S. Nitz. "Occurrence of Glycosidically Bound Flavor Compounds in Hops, Hop Products, and Beer," *Brewing Science* 59 (2006), 83-89.

Krofta, K., A. Mikyška, and D. Hašková. "Antioxidant Characteristics of Hops and Hop Products," *Journal of the Institute of Brewing* 114, no. 2 (2008), 160-166.

Laws, D.R.J. "A View on Aroma Hops," 1976 Annual Report of the Department of Hop Research, Wye College, 1977.

Lemmens, Gerard. "The Breeding and Parentage of Hop Varieties." 1998. Available at *www.brewerssupplygroup.com/FileCabinet/TheBreeding_Varieties.pdf*.

Levesque, John. *The Art of Brewing and Fermenting*. London: Thomas Hurst. 1836.

Lutz, A., K. Kammhuber, and E. Seigner. "New Trend in Hop Breeding at the Hop Research Center Hüll," *Brewing Science* 65, 2012.

Lutz, Henry. *Viticulture and Brewing in the Ancient Orient*. Leipzig, Germany: J.C. Hinkrichs'sche Buchhandlung, 1922. Reprint, Applewood Books, 2011.

McGorrin, Robert. "Character-impact Flavor Compounds," *Sensory-Directed Flavor Analysis*. Boca Raton, Fla.: CRC Press, 2007.

McPhee, John. *Oranges*. New York: Farrar Straus and Giroux, 1967.

Marshall, William. *The Rural Economy of the Southern Counties.* London: G. Nichol, J. Robinson, and J. Debrett, 1798.

Mathias, Peter. *The Brewing Industry in England, 1700-1830.* Cambridge, England: Cambridge University Press, 1959. Reprint, Gregg Revivals, 1993.

Meeker, Ezra. *The Busy Life of Eighty-Five Years of Ezra Meeker.* Seattle: Ezra Meeker, 1916.

_____. *Hop Culture in the United States.* Puyallup, Washington Territory: E. Meeker & Co., 1883.

Morimoto, M., T. Kishimoto, M. Kobayashi, N. Yako, A. Iida, A. Wanikawa, and Y. Kitagawa. "Effects of Bordeaux Mixture (Copper Sulfate) Treatment on Black Currant/Muscatlike Odors in Hops and Beer," *Journal of the American Society of Brewing Chemists* 68 (2010), 30-33.

Murakami, A., P. Darby, B. Javornik, M.S.S. Pais, E. Scigner, A. Lutz, and P. Svoboda. "Molecular Phylogeny of Wild Hops, *Humulus lupulus* L," *Heredity* 97 (2006), 66-74.

Myrick, Herbert. *The Hop: Its Culture and Cure, Marketing and Manufacture.* Springfield, Mass: Orange Judd Co., 1899.

Neve, R.A. *Hops.* London: Chapman and Hall, 1991.

Nickerson, G.B., and E.L. Van Engel. "Hop Aroma Profile and the Aroma Unit," *Journal of the American Society of Brewing Chemists* 50 (1992), 77-81.

Nielsen, Tom. "Dissecting Hop Aroma in Beer." Presentation at the Craft Brewers Conference, San Diego, 2008.

Ockert, Karl, ed. *MBAA Practical Handbook for the Specialty Brewer.* Vol. 1: *Raw Materials and Brewhouse Operations.* St. Paul, Minn.: Master Brewers Association of the Americas, 2006.

One Hundred Years of Brewing. Chicago, New York: H.S. Rich & Co., 1903. Reprint, Arno Press, 1974.

"100-year Birth Anniversary of Doc. dr. ing. Karel Osvald (*sic*)." *www.beer.cz/chmelar/international/a-stolet.html.*

Orwell, George. "Hop-picking," *New Statesman and Nation*, Oct. 17, 1931.

Osborne, Lawrence. *Accidental Connoisseur.* New York: North Point Press, 2004.

Patzak, J., V. Nesvadba, A. Henychova, and K. Krofta. "Assessment of the Genetic Diversity of Wild Hops (*Humulus lupulus* L.) in Europe Using Chemical and Molecular Markers," *Biochemical Systematics and Econology* 38 (2010), 136-145.

Patzak, J., V. Nesvadba, K. Krofta, A. Henychova, A. Marzoen, and K. Richards. "Evaluation of Genetic Variability of Wild Hops (*Humulus lupulus* L.) in Canada and the Caucasus Region by Chemical and Molecular Methods," *Genome* 53 (2010), 545-557.

Peacock, V., and M.L. Deinzer. "Chemistry of Hop Aroma in Beer," *Journal of the American Society of Brewing Chemists* 39 (1981), 136-141.

Peacock, Val. "Hop Chemistry 201." Presentation at the Craft Brewers Conference, Austin, Texas, 2007.

_____. "Percent Cohumolone in Hops: Effect on Bitterness, Utilization Rate, Foam Enhancement, and Rate of Beer Staling." Presentation at Master Brewers Association of the Americas Conference, Minneapolis, 2011.

_____. "Proper Handling, Shipping, and Storage of Hop Pellets." Presentation at Craft Brewers Conference, Chicago, 2010.

_____. "The Value of Linalool in Modeling Hop Aroma in Beer," *Master Brewers Association of the Americas Technical Quarterly* 47, vol. 4 (2010), 29-32.

Percival, John. "The Hops and Its English Varieties," *Journal of the Royal Agricultural Society of England* 62 (1901), 67-95.

Praet, T., F. Van Opstaele, B. Jaskula-Goiris, G. Aerts, and L. De Cooman. "Biotransformations of Hop-derived Aroma Compounds by *Saccharomyces cerevisiae* Upon Fermentation," *Cerevisia*, vol. 36 (2012), 125-132.

Pries, F., and W. Mitter. "The Re-discovery of First Wort Hopping," *Brauwelt* (1995), 310-311, 313-315.

Probasco, G., S. Varnum, J. Perrault, and D. Hysert. "Citra—A New Special Aroma Hop Variety," *Master Brewers Association of the Americas Technical Quarterly* 47, vol. 4 (2010), 17-22.

Proust, Marcel, trans. by Lydia Davis. *In Search of Lost Time, Vol. 1: Swann's Way*. New York: Penguin Group, 2003.

Salmon, E.S. "Notes on Hops," *Journal of the South-Eastern Agricultural College, Wye, Kent*, no. 42 (1938), 47-59.

Salmon, E.S. "Two New Hops: 'Brewers Favorite' and 'Brewers Gold,'" *Journal of the South-Eastern Agricultural College, Wye, Kent*, no. 34 (1934), 93-106.

Schmelzle, Annette. "The Beer Aroma Wheel," *Brewing Science* 62 (2009), 26-32.

Schönberger, C., and T. Kostelecky. "125th Anniversary Review: The Role of Hops in Brewing," *Journal of the Institute of Brewing* 117, no. 3 (2011), 259-267.

Schönberger, Christina. "Bitter Is Better," *Brewing Science* 59 (2006), 56-66.

_____. "Global Trends in Beer Bitterness," *Brauwelt International* 2011/1, 29-31.

_____. "Why Cohumulone Is Better Than Its Reputation," *Brauwelt International* 2009/III, 158-159.

Seefelder, S., H. Ehrmaier, G. Schweizer, and E. Seigner. "Genetic Diversity and Phylogenetic Relationships Among Accessions of Hop, *Humulus lupulus*, As Determined by Amplified Fragment Length Polymorphism Fingerprinting Compared With Pedigree Data," *Plant Breeding* 119, issue 3 (June 2000), 257-263.

Shellhammer, Thomas, ed. *Hop Flavor and Aroma: Proceedings of the 1st International Brewers Symposium.* St. Paul, Minn.: Master Brewers Association of the Americas and American Society of Brewing Chemists, 2009.

Shellhammer, Thomas. "Techniques for Measuring Bitterness in Beer." Presentation at Craft Brewers Conference, San Diego, 2012.

Shellhammer, Thomas, and Daniel Sharp. "Hops-related Research at Oregon State University." Presentation at Craft Brewers Conference, San Francisco, 2011.

Shellhammer, Thomas, D. Sharp, and P. Wolfe. "Oregon State University Hop Research Projects." Presentation at Craft Brewers Conference, San Diego, 2012.

Shepherd, Gordon M. *Neurogastronomy: How the Brain Creates Flavor and Why It Matters.* New York: Columbia University Press, 2012.

Simmonds, P.L. *Hops: Their Cultivation, Commerce, and Uses in Various Countries.* London: E. & F.N. Spon., 1877.

Smith, D.C. "Varietal Improvement in Hops," *Year Book of Agriculture*. Washington, D.C.: Government Printing Office, 1937, 1215-1241.

Southby, E.R. *A Systematic Handbook of Practical Brewing*. London: E.R. Southby, 1885.

Spinney, Laura. "You Smell Flowers, I Smell Stale Urine," *Scientific American* 304, issue 2 (2011), 26.

Stevenson, R.J., J. Prescott, and R. Boakes. "Confusing Tastes and Smells: How Odours Can Influence the Perception of Sweet and Sour Tastes," *Chemical Senses* 24 (1999), 627-635.

Stratton, Rev. J.Y. *Hops and Hop-Pickers*. London: Society for Promoting Christian Knowledge, 1883.

Sykes, W., and A. Ling. *The Principles and Practice of Brewing*. London: Charles Griffin & Co., 1907.

Takoi, K., Y. Itoga, K. Koie, T. Kosugi, M. Shimase, Y. Katayama, Y. Nakayama, and J. Watari. "The Contribution of Geraniol Metabolism to the Citrus Flavour of Beer: Synergy of Geraniol and ß-citronellol Under Coexistence With Excess Linalool," *Journal of the Institute of Brewing* 116, no. 3 (2010), 251-260.

Techakriengkrai, I., A. Paterson, B. Taidi, and J. Piggott. "Relationships of Sensory Bitterness in Lager Beers to Iso-Alpha Acid Contents," *Journal of the Institute of Brewing* 110, no. 1 (2004), 51-56.

Thausing, Julius, A. Schwarz, and A.H. Bauer. *Theory and Practice of the Preparation of Malt and the Fabrication of Beer*. Philadelphia: Henry Carey Baird & Co., 1882. Reprint, *BeerBooks.com*, 2007.

Toupin, Alice. *MOOK-SEE, MOXIE, MOXEE: The Enchanting Moxee Valley, Its History and Development*, 1970. Available at *www.evcea.org/evcea_about/Moxee.pdf*.

Trubeck, Amy. *The Taste of Place*. Berkeley: University of California Press, 2008.

Turin, Luca. *The Secret of Scent: Adventures in Perfume and the Science of Smell*. New York: Harper Perennial, 2007.

Unger, Richard. *Beer in the Middle Ages and Renaissance*. Philadelphia: University of Pennsylvania Press, 2004.

Van Opstaele, F., G. De Rouck, J. De Clippeleer, G. Aerts, and L. Cooman. "Analytical and Sensory Assessment of Hoppy Aroma and Bitterness of Conventionally Hopped and Advance Hopped Pilsner Beers," *Institute of Brewing & Distilling* 116, no. 4 (2010), 445-458.

Van Opstaele, F., Y. Borremans, A. Van Holle, J. Van Nieuwenhove, P. De Paepe, D. Naudts, D. De Keukeleire, G. Aerts, and L. De Cooman. "Fingerprinting of Hop Oil Constituents and Sensory Evaluation of the Essential Oil of Hop Pellets From Pure Hop Varieties and Single-hop Beers Derived Thereof." Poster presented at Tenth Trends in Brewing, Ghent, Belgium. 2012.

Vogel, E., F. Schwaiger, H. Leonhardt, and J.A. Merten. *The Practical Brewer: A Manual for the Brewing Industry*. St. Louis: Master Brewers Association of America, 1946.

Vogel, M. *On Beer: A Statistical Sketch*. London: Tribner & Co., 1874.

Wahl, Robert, and Max Henius. *American Handy Book of the Brewing, Malting, and Auxiliary Trades*. Chicago: Wahl & Henius, 1901.

Webb, Tim, and Stephen Beaumont. *The World Beer Atlas*. New York: Sterling Epicure, 2012.

Whittock, S., A. Price, N. Davies, and A. Koutoulis. "Growing Beer Flavour—A Hop Grower's Perspective." Presentation at Institute of Brewing and Distilling Asia Pacific Section Convention, Melbourne, Australia, 2012.

Wilson, D. Gay. "Plant Remains From the Graveney Boat and the Early History of *Humulus lupulus* L. in Europe," *New Phystol* 75 (1975), 627-648.

Wright, W.E. *A Handy Book for Brewers*. London: Lockwood, 1897.

Index

Entries in **boldface** refer to photos and illustrations

xanthohumol, 47, 71, 179

X-114, 79-80

Yakima Chief Ranches, 83, 158. *See also* Lemmens, Gerard

Yakima Valley, 88, 93, 94, 102, 104, 107. *See also* Washington
  historically, 117-118
  powdery mildew in, 123
  2011 harvest in, 98

yeast, 26, 217

Yeoman, 171

Žatec region and town, 2, 59, 101, **color plate 4.** *See also* Czech Republic

Zeiner, Hans, 94, 95, 99

Zeus, 167, 173

Zimmerman, Chuck, 82

ZYTHOS, 138